Springer Theses

Recognizing Outstanding Ph.D. Research

Aims and Scope

The series "Springer Theses" brings together a selection of the very best Ph.D. theses from around the world and across the physical sciences. Nominated and endorsed by two recognized specialists, each published volume has been selected for its scientific excellence and the high impact of its contents for the pertinent field of research. For greater accessibility to non-specialists, the published versions include an extended introduction, as well as a foreword by the student's supervisor explaining the special relevance of the work for the field. As a whole, the series will provide a valuable resource both for newcomers to the research fields described, and for other scientists seeking detailed background information on special questions. Finally, it provides an accredited documentation of the valuable contributions made by today's younger generation of scientists.

Theses are accepted into the series by invited nomination only and must fulfill all of the following criteria

- They must be written in good English.
- The topic should fall within the confines of Chemistry, Physics, Earth Sciences, Engineering and related interdisciplinary fields such as Materials, Nanoscience, Chemical Engineering, Complex Systems and Biophysics.
- The work reported in the thesis must represent a significant scientific advance.
- If the thesis includes previously published material, permission to reproduce this must be gained from the respective copyright holder.
- They must have been examined and passed during the 12 months prior to nomination.
- Each thesis should include a foreword by the supervisor outlining the significance of its content.
- The theses should have a clearly defined structure including an introduction accessible to scientists not expert in that particular field.

More information about this series at http://www.springer.com/series/8790

Claudia Aguilera-Gómez

Explaining Lithium Enriched Red Giant Branch Stars

Doctoral Thesis accepted by
Pontifical Catholic University of Chile, Santiago, Chile

Springer

Claudia Aguilera-Gómez
Departamento de Ciencias Físicas
Facultad de Ciencias Exactas
Universidad Andrés Bello
Las Condes, RM – Santiago, Chile

ISSN 2190-5053 ISSN 2190-5061 (electronic)
Springer Theses
ISBN 978-3-030-02582-3 ISBN 978-3-030-02583-0 (eBook)
https://doi.org/10.1007/978-3-030-02583-0

Library of Congress Control Number: 2018960168

This Springer imprint is published by the registered company Springer Nature Switzerland AG
The registered company address is: Gewerbestrasse 11, 6330 Cham, Switzerland

Supervisor's Foreword

The current theory of stellar structure and evolution was born and grew up along with the revolutions of modern and quantum physics of the early twentieth century. It is arguably one of the most successful physical frameworks of modern astrophysics, explaining the inner workings of stars of all sizes and chemical compositions to a remarkably high level of agreement between theory and observation. The model not only accounts for the detailed interior of stars at any given time (their structure, state of the matter, energy generation, energy transport, internal dynamics) but also successfully explains precisely why and how these properties change in time almost since stars are born until their ultimate fate in a universe of a finite age. Moreover, when the machinery of the model is applied to large collections of stars, we are able to understand the behavior of the stellar populations that make up and drive the evolution of the variety of galaxies whose studies, in turn, teach us about cosmology and even provide tests of some of the fundamental physics on which the theory itself rests. Today, our understanding of stellar interiors is being further boosted by the maturity of the tool of asteroseismology, which provides a new ability to peer into the structure and dynamics of stellar interiors. At the same time, the asteroseismic advances on stellar astrophysics, along with the present astrometric revolution being enabled by the *Gaia* mission and various large spectroscopic surveys, are leading us today into a new era in the understanding of the formation, structure, and evolution of the Milky Way galaxy as a whole. Modern stellar astrophysics, therefore, is an extremely robust construction that keeps improving and upon which new and larger astrophysical enterprises rest. In this context, long-standing unsolved problems in the area are few, and any new advances that offer truly novel insight on those problems are, not surprisingly, extremely rare. Claudia Aguilera-Gómez's PhD thesis work is one of these rarities.

The existence of a small but ubiquitous fraction of red giant stars with large amounts of lithium in their surfaces is among those few old problems in stellar astrophysics that have eluded an explanation for a few decades now. The problem is made the more difficult due to the fact that the physics of lithium is relatively simple (and therefore all available options to solve the problem have been explored already, or seemingly so), and most observational data involving it are satisfactorily explained today with standard astrophysics. Lithium is consumed (burned) via proton capture at relatively low temperatures for stellar interior standards, and this fragility is exploited in stellar astrophysics to obtain a probe, a thermometer, for the deep interiors of stars, i.e., locations that we are prevented from seeing by direct means because stars are such opaque objects to the passage of light. Since we know for a fact that lithium is burned at temperatures above 1.5–2 million degrees, all that we know after almost a hundred years of research in stellar astrophysics indicates almost without a doubt that when a low-mass star like the Sun gets old enough and grows in size to become a red giant, the lithium present in its surface should (1) first dissolve, because the outer convective envelope of the star is getting a lot larger, and (2) burn and disappear almost completely, because the deeper layers of this envelope get hotter than the burning temperature of lithium. However, while most red giants seem to follow this expectation, it happens that wherever we look in the Milky Way and around it, we find some red giants with large amounts of lithium. Early since their discovery, and this fact is not anecdotal in this story, lithium-rich giants were defined to be those objects with more than three times more lithium in their surfaces than in the Sun's. Even with observational instrumentation and data improving with the years, and models becoming more and more sophisticated, the decades have passed and the problem remains there, with basically no new insights on where to go.

Known potential solutions to the problem could be separated in two classes, depending on whether the excess lithium was produced within the star itself or was acquired from an external source. Claudia chose to approach the problem from the latter perspective and studied the consequences of the ingestion (by the star) of objects such as planets or brown dwarfs at some point during the star's growth into the red giant branch—for example, Mercury, Venus, and possibly the Earth will all be ingested by the Sun some 4.5 billion years from now when our star becomes a red giant. The idea was simple: since planets and brown dwarfs contain some lithium in them, when they are ingested by a growing red giant this lithium is incorporated by it, and thus the red giant might appear as containing more lithium than expected from standard theory. While not new, this alternative had never been quantitatively explored in all its consequences, and Claudia's thesis showed that about half of all the known lithium-rich red giants can be explained by this process, thus effectively solving one half of a problem about three decades old.

While that would already amount to a successful PhD thesis, Claudia did not stop there and went ahead to perform a clever population-level study in which she explored how a large number of red giants would look like if they all ingested at least one planet. By thoroughly accounting for all possible initial conditions consistent with observations, both for the star and for the planetary systems around them, Claudia obtained the crucial insight that makes her thesis work truly game-changing. She realized that the observational phenomenology that astronomers had been using as the empirical evidence informing the problem of lithium-rich giants had been a misleading one. For decades, the fixed threshold used for identifying lithium-rich giants had been classifying them wrongly, and some stars that appeared to be lithium-rich actually were normal stars (for their mass), while other stars that appeared to be normal in lithium were actually lithium-rich (given their mass). The result was that efforts to address the problem were narrowly focused on explaining the largest, most extreme lithium enhancements, not realizing that the phenomenology of the problem was not restricted to those extreme stars. Under the light of this truly new perspective, it is no wonder then that a solution has been so difficult to find and has puzzled astronomers for so long.

Today, a still important part of the problem of lithium-rich giants remains to be solved, but, thanks to Claudia's PhD thesis, at least we have now learned how to fulfill a fundamental condition for any hope of a solution, i.e., how to adequately inform ourselves of the real phenomenology behind the problem. Thanks to Claudia's PhD thesis, for example, we now know how to adequately select samples of stars aimed to tackle specific questions related to the problem, and this important insight holds the key for the dismantling of this 30+ year-old puzzle into its most fundamental pieces.

Institute of Astrophysics
Pontificia Universidad Católica de Chile
Santiago de Chile
Chile

Julio Chanamé

Acknowledgements

This book is based on my doctoral thesis, presented at the Astrophysics Institute at Pontificia Universidad Católica de Chile in June 2017.

I have many people to thank for contributing either directly or indirectly to this thesis and my research. I would like to thank Julio Chanamé, for supporting my research and guiding my progress throughout the doctorate. He provided valuable academic advice, fruitful discussions, and personal support when I needed it. I would also like to thank M. Catelan, M. Zoccali, and M. Pinsonneault for their advice and for proofreading this thesis.

To all of the professors at the Astrophysics Institute, a big thank you for sharing your knowledge and for all the scientific discussions. I am also grateful to all the non-scientific staff at the Astrophysics Institute and MAS, the heroes that helped me on a daily basis with administrative and IT issues.

I would also like to acknowledge my collaborators. M. Pinsonneault, J. Carlberg, I. Ramirez, and M. Jones. Thank you all for the suggestions, support, ideas, and hard work. I also thank L. Monaco and L. Sbordone for the debates and interesting scientific conversations. Also, thanks to M. Pinsonneault and the people in the Astronomy Department at The Ohio State University for the warm welcome during my short stay.

I acknowledge support from CONICYT FONDECYT Postdoctorado 2018 proyecto número 3180668. I am also grateful to the following for financial support: CONICYT-PCHA Doctorado Nacional 2013-21130353; the Chilean Ministry for the Economy, Development, and Tourism's Programa Iniciativa Científica Milenio, through grant IC120009 awarded to the Millennium Institute of Astrophysics (MAS); PFB-06 Centro de Astronomia y Tecnologias Afines; and the Vicerrectoría de Investigación Pontificia Universidad Católica de Chile. Also thanks to the people behind the tools and catalogs I used during this research: TOPCAT, IRAF, MOOG, NASA ADS Service, the Extrasolar Planets Encyclopaedia, and the VizieR catalog access tool, CDS, Strasbourg, France.

On a more personal note, thanks to all of my friends. Special thanks to Sergio, Pedro, María Paz, and Felipe, for all those days finishing homework, discussing each others research, or talking about unrelated nonscientific nerdy stuff. Also, thank you Newton, Morgana, and Leia.

I am forever grateful to my family for their unconditional support. Without them, I would not be the person I am today. Thanks to my parents Claudio and Viviana for their love, encouragement, valuable advice, and for the opportunities they gave me. They always believe in me, even when I don't. Thanks to Lucas for being my only ally in the evil lair of health-care providers that is our home. His funny remarks and silent support were there whenever I needed them. Thanks to Andrés, my constant source of love, encouragement, protection, and positivity. His patience and support in each step of this astronomical journey always remind me of what is truly important in life.

Contents

Parts of this thesis have been published in the following journal articles:

1. Aguilera-Gómez, C., Chanamé, J., Pinsonneault, M. & Carlberg, J. K. On Lithium-rich Red Giants. I. Engulfment of Substellar Companions, ApJ, 829, 127. http://doi.org/10.3847/0004-637X/829/2/127 (2016).
2. Aguilera-Gómez, C., Chanamé, J., Pinsonneault, M. & Carlberg, J. K. On Lithium-rich Red Giants: Engulfment on the Giant Branch of Trumpler 20, ApJL, 833, 24. http://doi.org/10.3847/2041-8213/833/2/L24 (2016).
3. Aguilera-Gómez, C., Ramírez, I. & Chanamé J. Lithium abundance patterns of late-F stars: an in-depth analysis of the lithium desert, A&A, 614, 55. http://doi.org/10.1051/0004-6361/201732209 (2018).

Chapter 1
Introduction

Stars are one of the building blocks of the Universe. Most of the information that we receive from our Galaxy or others comes in form of starlight, which then must be interpreted by using physical theories, such as the stellar evolutionary theory developed during the last century. Studying the composition, evolution, and nature of stars leads us to understand the formation and evolution of planetary systems, as our own, the chemical evolution of the Universe and the production of different elements, and the galaxies and populations where they reside. Also, stars are ideal laboratories to test new (and old) physics, such as rotation, element production, magnetic fields, or even physics at extreme conditions in stellar remnants, among many others.

In particular, studying observational anomalies in astronomy that are not explained with physics currently implemented in models allows understanding of what else is needed to create a fully comprehensive picture of our Universe.

In this introduction, I will briefly summarize modern stellar evolutionary models (Sect. 1.1). Then, in Sect. 1.2, I will introduce the light element lithium and how its abundance can be used to increase our understanding of stars, extending this to the specific problem that I study in this work in Sect. 1.3, lithium-enhanced red giants.

1.1 Stellar Evolutionary Models

Stellar evolutionary models are simplified descriptions of stars that are constructed to describe them and make predictions. The advancement of technology and processing power in computers has allowed to solve the coupled equations that define stellar structure and evolution (Kippenhahn et al. 1967). At the same time, a deep theoretical understanding of the physics of stars was also needed to first develop these models and is still needed today to understand them and use them properly (Clayton 1968; Kippenhahn and Weigert 1990). It has taken over a century

C. Aguilera-Gómez, *Explaining Lithium Enriched Red Giant Branch Stars*, Springer Theses, https://doi.org/10.1007/978-3-030-02583-0_1

for astronomers to build the physical models that describe stars reaching their current stage of development.

The standard models of stellar evolution solve the four basic equations of stellar structure, i.e., equations of hydrostatic equilibrium, mass and energy conservation, and energy transport, assuming spherical symmetry. The input physics included in the models accounts for convective, conductive, and radiative energy transport, nuclear reactions and energy production, the equation of state, opacities, microscopic diffusion, and more. Some physics is usually neglected, like the presence of magnetic fields or rotation. Some information about these processes and how they could affect the stellar evolution can be found in Tassoul (2007).

The equations of stellar evolution in spherical symmetry depend only on the distance to the center of the star and the time in the evolution, both independent quantities. As such, the problem of solving these equations is one-dimensional. During the evolution of the star, its radius can change significantly, but the mass remains constant during the main sequence and only changes in more evolved stages of the evolution. Thus, in practice, the equations of stellar evolution are solved by using the mass of each shell as an independent variable. In conclusion, the pressure, temperature, radius, and luminosity of the star are solved at a given time of the evolution and as a function of the mass in each shell of the star.

In spite of the simplifications, standard stellar models have been able to reproduce observations, such as the color-magnitude diagrams of stellar populations like clusters, and they can accurately describe several properties of the Sun (Serenelli 2016). In particular, there is a good agreement between predictions from these models for the internal stellar structure of the Sun and helioseismology. The stellar evolutionary models were initially created to reproduce the parameters of the Sun, and current models are still calibrated and adjusted using the observed and measured parameters of our star. The agreement between models and observation is encouraging, but there are indications that the standard model does not include all the necessary physics. Based on the discrepancies of the standard models with observations, we can test additional processes acting inside stars.

One of the effects of noncanonical physics, which includes rotation, planet-star interactions, and non-convective mixing, is to significantly alter the light element abundance of stars in all evolutionary phases. Thus, to explain the abundance patterns found in the Sun and other stars in the Galaxy, more recent models have started to include new physics. Then, as we will see throughout this work, the stars that do not display the expected behavior, according to standard models, are critical to complement our current knowledge on stellar evolution and the general physics dominating our Universe.

1.2 Lithium in Astrophysics

Lithium (Li) is a light element often used to study different processes taking place inside stars. It was produced shortly after the Big Bang, but it is easily destroyed by proton capture at the temperatures and densities found inside main sequence

(MS) stars. Thus, the survival of this element depends on the conditions of the environment. Lithium is also created in some specific environments. Although Li is scarce, its abundance in stars can lead us to understand the production of elements in the early Universe, minutes after the Big Bang, the chemical enrichment of our Milky Way and other stellar populations, and stellar evolution.

There are two different stable lithium isotopes found in nature. Lithium-6 is far less abundant than Lithium-7, which is the focus of this thesis. Throughout this work, we will refer to the lithium-7 isotope simply as lithium or Li and specifically mention when we are considering its isotope.

I will summarize here how the Li abundance is used in astrophysics, outlining its main characteristics and some of the different problems it presents.

1.2.1 Big Bang Nucleosynthesis and the Lithium Problem

The standard model of Big Bang Nucleosynthesis (BBN) describes the production of light elements in the early Universe. Lithium-7 is one of the few elements that are produced in these first minutes after the Big Bang when the Universe was hot enough, along with deuterium, helium-4, and helium-3 (Wagoner et al. 1967), where the primordial abundance of these elements depends on the baryon density in the early Universe. More baryons in the early Universe would imply more reactions and more formation of deuterium, which is needed to produce heavier elements.

The baryon density can be obtained from observations of the cosmic microwave background (CMB) in the form of the baryon-to-photon ratio, thus allowing to predict the primordial abundance of elements produced in the standard BBN model. With observations of the Wilkinson Microwave Anisotropy Probe (WMAP, Komatsu et al. 2011), a primordial abundance of Li $A(Li) = 2.72$ is expected (Coc et al. 2012). A similar value of primordial Li is obtained with the baryon density from the Planck Mission (Planck Collaboration et al. 2016). The abundance of Li is usually measured on a scale where the hydrogen abundance is 12:

$$A(\mathrm{Li}) = \log(\mathrm{N_{Li}}/\mathrm{N_H}) + 12.00, \qquad (1.1)$$

where N_x is the number of atoms of element "x".

Observations of the most primitive astrophysical sites possible allow measuring the primordial abundances of deuterium, helium-3, and helium-4 that are compared with predictions from standard BBN calculations and baryon-to-photon ratios from cosmic microwave background measurements. A good agreement is found for all these elements being this one of the cornerstones of the Big Bang theory. The first two panels of Fig. 1.1 show how the abundance of these elements changes with baryon-to-photon ratio. The vertical lines are the value of baryon density found by WMAP and Planck. The abundances predicted by standard BBN (the intersection between blue lines and the vertical bands in each panel) agree well with observations (green horizontal lines).

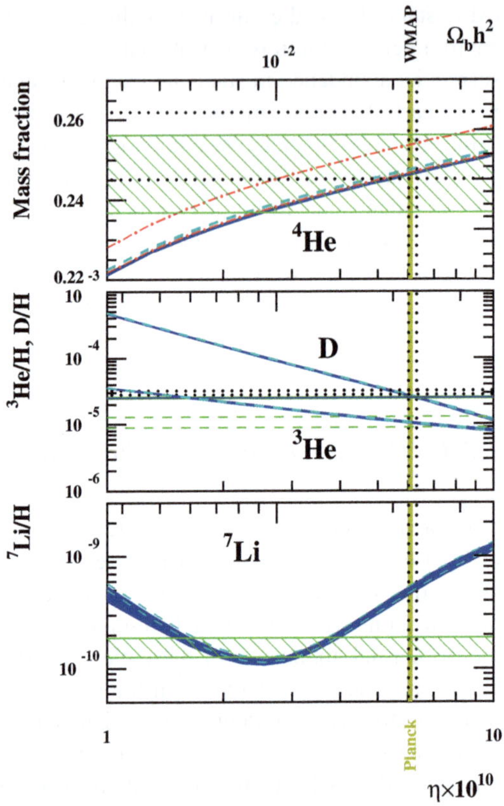

Fig. 1.1 Figure from Coc et al. (2014). ^{4}He, ^{3}He, D, and ^{7}Li abundances as a function of baryon-to-photon ratio (η) calculated with an updated full network of nuclear reactions (dark blue, Coc et al. 2014) or reduced network (light blue dashed lines, Coc and Vangioni 2010). Vertical bands correspond to the WMAP (dotted black) and Planck (yellow solid band) baryon densities. The horizontal bands represent the adopted observed values of each element abundance. We see an agreement in the top two panels between observations and predicted abundances using BBN and either WMAP or Planck baryonic densities, while a strong disagreement is seen in the bottom panel for lithium. Credit: Coc et al. 2014, *Standard big bang nucleosynthesis and primordial CNO abundances after Planck*, JCAP, 10, 050

In contrast, there is a discrepancy between the BBN predictions of Li abundance and what is measured in old, metal-poor main sequence stars, interpreted to represent the primordial abundance. Spite and Spite (1982) showed that the Li abundance of these stars is constant (the so-called *Spite Plateau*), a result that has been recovered by other authors, confirming an abundance of A(Li) $\sim$ 2.0–2.2 (Asplund et al. 2006; Bonifacio et al. 2007; Sbordone et al. 2010). This abundance is over 3 times lower than the predictions from BBN. This can be observed in the bottom panel of Fig. 1.1, where the green horizontal band is below the predicted value by CMB plus BBN. Moreover, an additional problem is found when studying more metal-poor stars that have consistently smaller Li abundances as they decrease their metal content, no longer forming a plateau (Ryan et al. 1996; Asplund et al. 2006; Sbordone et al. 2010).

This discrepancy is the "Lithium problem" (for a review, see Fields 2011). Possible solutions to this problem have been suggested, but there is not a definitive explanation yet. On one hand, changes in nuclear physics could alter the reactions during the BBN and decrease the predicted Li abundances from cosmic microwave background measurements (e.g., Boyd et al. 2010; Cyburt and Pospelov 2012). Also, solutions beyond the standard model of big bang nucleosynthesis are considered (for a review, see Pospelov and Pradler 2010). On the other hand, a different set of solutions for the Li problem suggests that the Li abundance measured in stars is not representative of the primordial Li abundance predicted by the BBN models. This set of explanations focus on different Li depletion mechanisms acting in halo stars (which we will discuss in more detail in Sect. 1.2.3) or in the complete Li depletion during the pre-MS and later accretion from residual discs to restore it (Fu et al. 2015).

1.2.2 Chemical Enrichment of the Galaxy

Despite the disagreement between predictions from BBN and measured Li abundance for metal-poor stars in the Galaxy, it is known that some kind of lithium enrichment is needed afterward to reach the values found in meteorites (A(Li)=3.26, Lodders et al. 2009) that resemble the composition of the solar nebula when the Sun was forming. Thus, the Li abundance can also be used to study the chemical enrichment of the Milky Way. Li can be produced in the interior of low-mass stars during their asymptotic giant branch phase (Sackmann and Boothroyd 1992). In this advanced stage of evolution, the inert core of the star is surrounded by two nuclear burning layers, one of helium and an outer layer of hydrogen, and a large convective envelope. Li es produced in convective conditions for these stars. Additionally, it has been suggested that stellar interiors could be an important site of Li production earlier in the life of the star, during its red giant branch phase (Sackmann and Boothroyd 1999), although this possibility is still under debate. Moreover, Li can also be produced in the Galaxy by cosmic ray spallation (Reeves 1970), core collapse supernovae (Woosley et al. 1990), and novae (Arnould and Norgaard 1975; Tajitsu et al. 2015).

Knowing the common sites of Li production is not enough to understand the chemical evolution of the Galaxy, given that the yields from these processes fail to match the observed Li abundances of the Galaxy (e.g., Romano et al. 2001; Travaglio et al. 2001; Prantzos 2012).

Building models of Li enrichment in the Milky Way is not only complicated by the unknown processes that modify the abundance in stellar interiors (see Sect. 1.2.3) but also because stellar ages are usually very uncertain and particularly hard to measure, and thus, it is not known how Li evolves with age. This makes it hard to compare models with observations. A commonly used approach to overcome these obstacles is to use the metallicity as a proxy of age and constrain the abundance of the interstellar medium when stars are formed. However, as Li is depleted at stellar interiors and can even be produced at some phases of stellar evolution, there is a large spread of Li with any metallicity. Further assumptions and selection of stars is needed to study the chemical enrichment, e.g., selecting only hotter star that have smaller convective envelopes or considering only the upper envelope of the Li distribution that is supposed to represent stars with no depletion and, as such, that are more representative of the interstellar medium (e.g., Delgado Mena et al. 2015; Guiglion et al. 2016).

1.2.3 Lithium in Stars

The Li abundance measured in meteorites is used as an indicator of the solar abundance at formation. The predictions of surface Li abundance of stars are made based on standard stellar models (Iben 1967). In low-mass stars with convective envelopes, these predictions indicate that Li decreases only when Li is burned in convective conditions. In other words, the temperature in the base of the convective zone must be higher than the temperature of Li burning ($T \gtrsim 2.5 \times 10^6$ K), which should only happen during the pre-main sequence.

Very low-mass stars that are fully convective burn in Myr scales all Li once they reach the needed temperature. For stars of higher masses, the convective envelope retreats at the end of the pre-MS, decreasing its temperature and finishing with the phase of Li depletion. The exact abundance of Li after the pre-MS also depends on the metallicity of the star, because higher metallicities imply larger envelopes and higher depletion factors. Thus, canonical stellar evolution predicts a certain Li abundance given the same mass and metallicity of stars.

No further surface depletion of Li is expected during the MS since the convective envelope remains at a temperature lower than that needed to burn Li. Confirming these theoretical predictions, helioseismic measurements (Christensen-Dalsgaard et al. 1991) indicate that the temperatures and densities needed to burn lithium inside stars are located below the convective envelope. Then, we should expect no depletion during the MS for the Sun. However, the solar photospheric Li abundance measured now of A(Li) = 1.05 (Asplund et al. 2009) is 2 dex lower than the value found in meteorites, marking a strong difference with predictions from canonical stellar evolution. Moreover, predictions from canonical stellar evolution indicate that there should be no dispersion of Li abundance for stars with similar atmospheric parameters, different from what is observed in MS stars (e.g., Lambert and Reddy 2004), where stars of fixed mass, metallicity, and age can have different Li abundances. Given that standard stellar models are not always accurate when describing the observed patterns in stars, there must be processes acting inside stellar interiors that are not considered in the models. In particular, the discrepancy between meteoritic and photospheric Li evidences a nonstandard Li depletion process taking place in stellar interior during the MS phase.

Some of the processes that could produce a depletion of Li during the MS are rotational mixing (Pinsonneault et al. 1992; Eggenberger et al. 2010), diffusion (Michaud 1986), mass loss (Swenson and Faulkner 1992), gravity waves (Montalbán and Schatzman 2000; Charbonnel and Talon 2005), and overshooting (Xiong and Deng 2009), among others.

This situation is different from what is found for other elements that have a similar abundance in both meteorites and the solar photosphere. Other exceptions are beryllium and boron, which are also underabundant in the Sun, and elements like hydrogen, helium, carbon, nitrogen, and oxygen that are produced in stellar interiors and, as such, are less abundant in meteorites (Lodders et al. 2009).

Li depletion in the Sun and solar analogs (stars similar to the Sun) is a yet unsolved problem. However, stars with Li-depleted surfaces are not restricted to that specific temperature range. Boesgaard and Tripicco (1986) studied the Li abundance pattern in the Hyades open cluster, finding that F-dwarfs with temperatures around 6600 K had large Li depletion factors when compared to stars either 300 K hotter or cooler. This well-known pattern known as the lithium dip affects stars of masses between 1.0–1.5 $M_{\odot}$, a location that changes with metallicity. This pattern has been found in other clusters of similar ages and older, such as M34 (Jones et al. 1997), NGC 752 (Hobbs and Pilachowski 1986), and M67 (Balachandran 1995), but it is not present in stars of younger clusters (e.g., the Pleiades, Boesgaard et al. 1988), pointing toward a depletion mechanism acting during the MS phase of stellar evolution. Although this feature is harder to distinguish in field stars, given the mix of different ages, masses, and metallicities, it has been confirmed to exist in this environment as well (Randich et al. 1999; Lambert and Reddy 2004). The exact depletion mechanism hidden under the Li dip feature is unknown; however, any mechanism invoked as an explanation must consider how the position of the lithium dip can change depending on the mass, metallicity, or age of the stars (e.g., Delgado Mena et al. 2015).

Hotter than solar analogs, but cooler than stars found in the Li dip, we found stars with temperatures $T_{\rm eff} \gtrsim 5900$ K. For these, Chen et al. (2001) measured either high or very low Li abundances, finding no stars with intermediate Li abundances. The empty area in the A(Li) $-$ $T_{\rm eff}$ diagram is the *lithium desert*. Based on the tight correlation between mass and metallicity present in stars with low Li abundances, a behavior that is observed for stars in the Li dip, Chen et al. (2001) proposed that stars in the low Li side of the desert could be connected to the Li dip. However, the exact same correlation between mass and metallicity was also found by Ramírez et al. (2012) in stars with high Li abundances and thus was not exclusive to the depleted stars. Other mechanisms of Li depletion acting during the MS or subgiant phases of stellar evolution could also produce a similar feature in Li.

The discovery of exoplanets orbiting stars also adds complexity to correlations of different stellar parameters that may be modified by the planets, including the Li abundance. All the effects of planets on the angular momentum of the star could eventually also impact their Li abundances. During the pre-main sequence, the star could interact with the protoplanetary disc, enhancing the rotationally induced mixing in the star, a process that decreases the Li abundance (Bouvier 2008). During the life of the star, the migration of planets can affect their angular momentum and Li (Castro et al. 2009). These works suggest that main sequence stars that are planet hosts could have lower Li abundance than stars without planets orbiting them. The scientific community is currently divided on this topic. Some works do find observational evidence of an enhanced Li depletion in planet hosts (Israelian et al. 2004; Gonzalez 2008; Israelian et al. 2009; Delgado Mena et al. 2014; Figueira et al. 2014; Gonzalez 2015), while others found no correlation or suggest that there are biases that justify the presence of these correlations, so they do not have physical meaning (Ryan 2000; Luck and Heiter 2006; Baumann et al. 2010; Ghezzi et al. 2010; Ramírez et al. 2012; Carlos et al. 2016).

All of the problems described in this section show that Li rarely behaves how it is supposed to, considering standard models. It is a difficult element to study, as its abundance depends on mass, effective temperature, metallicity, age, evolutionary stage, and even additional variables that we have not been able to understand completely (e.g., rotation, mixing, and mass transfer). An added difficulty when working with this element is that it is scarce, and as such, it is difficult to measure in the spectra of stars, usually needing high resolution and high signal-to-noise spectra to even distinguish the Li doublet located at 6707.8 Å.

1.3 Red Giant Branch Stars

Low-mass stars undergo a series of changes when they enter their post-MS evolution. The central hydrogen (H) fusion that has supported the star during most of its lifetime in the MS has ended. The helium core is now inert, and gravity will cause the core to contract, increasing its temperature and producing energy. When the core reaches a critical temperature, the star ignites H in a thin shell around the core, burning it steadily. Due to the larger pressure that the outer layers of the star experience, they will expand significantly. This increase in the radius is accompanied by a decrease in effective temperature. As the outer layers of the star cool, they become fully convective, a process that enables a more efficient transport of energy and thus increasing abruptly the luminosity of the star, becoming a red giant branch (RGB) star. This full process is illustrated in the Hertzsprung-Russell diagram of Fig. 1.2, where I show both the luminosity (L) and surface gravity (g, which relates to the mass and radius of the star) as a function of the effective temperature (T_{eff}) of the star. Even if both of these quantities show a similar post-MS evolution, the surface gravity can be directly measured from stars, while to obtain the luminosity, additional information is needed about the distance.

Once the star reaches the red giant branch, it also suffers changes in its surface abundances. The surface convection zone deepens, mixing the nuclear-processed material from the interior with material from the surface of the star, in the process known as first dredge up (FDU). This process homogenizes the outer layers of the star and can carry the byproducts of H-burning outward. Examples of changes in element abundances during this process are the increase of N, the decrease of C, and the decrease in carbon isotopic ratio ($^{12}C/^{13}C$).

In particular, we have seen that Li has a complicated history during the MS (see Sect. 1.2). Its abundance at the end of the MS phase depends on several factors like the possibility of mixing and the star's mass. And although the interior of the stars is always devoided of Li, we can find stars entering their post-MS evolution with almost no surface depletion from the meteoritic value and stars like the Sun that have already decreased their abundance by a large factor.

During the FDU, the Li-depleted interior is mixed with the relatively rich exterior, decreasing the abundance. The Li of a star that begins its RGB with the solar meteoritic abundance should, therefore, be ~ 1.8 dex lower than the MS turn-

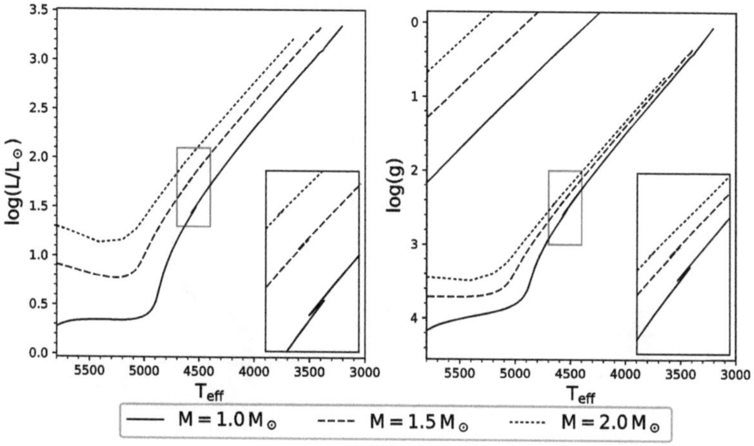

Fig. 1.2 Left panel: Hertzsprung-Russell diagram showing stellar evolutionary tracks calculated with the YREC stellar evolutionary code for stars of masses 1.0 M$_\odot$, 1.5 M$_\odot$, and 2.0 M$_\odot$ and solar metallicity. All of these tracks reach the end of the red giant branch. The inset shows the luminosity function bump when the advancing H-burning shell erases the chemical discontinuity left by the FDU. Right panel: Same tracks as in the left panel but instead of luminosity, these show log g, a property that is directly observable for stars

off value. This sets an upper limit of A(Li) $=$ 1.5 for what the Li abundance is expected to be in giant stars. In general, we expect much lower Li in the RGB phase, depending on the stellar mass and metallicity and also considering Li burning during the FDU.

In the meantime, the H-burning shell is progressing outward, and, eventually, the surface abundances of the star change again. This is a process that only happens for low-mass stars when the H-burning shell crosses the discontinuity in the μ-profile that was left in the star after the FDU. This is the luminosity function bump or RGB bump (see the inset in Fig. 1.2). Some authors (e.g., Denissenkov and VandenBerg 2003) suggests that before the H-burning shell erases this barrier in mean molecular weight, the important μ gradients do not allow mixing to proceed, and thus material from the interior of the star cannot reach the convective envelope. Chanamé et al. (2005) instead calculated rotational mixing inside the star, finding that the mixing is inefficient before the RGB and not because of the barrier in molecular weight. Regardless of the reason why no surface abundance changes are present before the RGB bump, there is strong observational evidence indicating that there is extra-mixing after this moment in evolution that changes the surface abundance of elements like Li, C, N, O, and the carbon isotopic ratio (e.g., Spite et al. 2005; Gratton et al. 2000; Shetrone 2003). This process, known as canonical extra-mixing (Denissenkov and VandenBerg 2003) would decrease the surface Li abundance of stars (Lind et al. 2009b; Palmerini et al. 2011). Studies on globular clusters have suggested that the deep mixing is more efficient in low-metallicity stars, based on the abundance of certain light elements (e.g., carbon Martell et al.

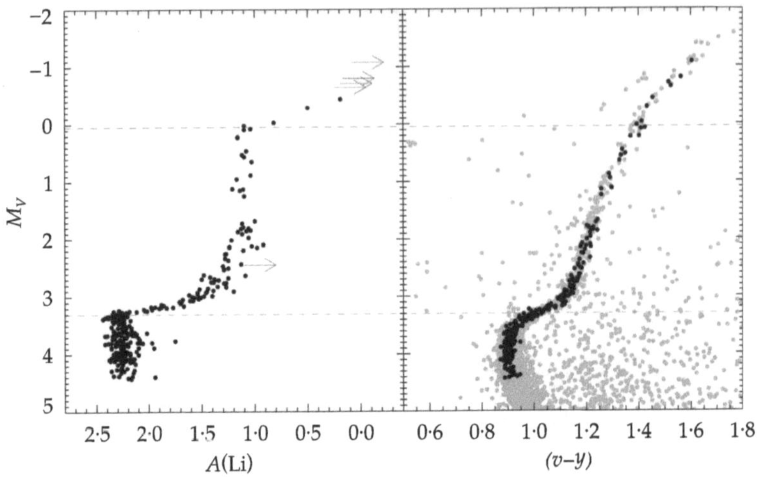

Fig. 1.3 Figure 5 from Lind et al. (2009b). Left panel: Absolute visual magnitude and non-LTE Li abundances in the sample of stars members of the metal-poor globular cluster NGC6397. Here arrows are Li upper limits. Right panel: color-magnitude diagram of the cluster. In both panels, the dashed lines show the start of Li decrease in two different points of stellar evolution, the first dredge up, starting at $M_V \sim 3.3$, and the luminosity function bump at $M_V \sim 0$. Credit: Lind et al. (2009b), A&A, 503, 545, reproduced with permission ©ESO

2008), but a better sampling of different stellar properties is needed to conclude and relate the dependencies on these parameters with the actual physical mechanism behind the mixing.

In Fig. 1.3 (from Lind et al. 2009b), we see the Li abundance pattern of the metal-poor globular cluster NGC 6397 (left panel) and its color-magnitude diagram (right panel). This figure clearly illustrates the points in stellar evolution where there is a decrease in the surface Li abundance. The first episode is the FDU, the deepening of the convective layer, where Li is diluted, starting at $M_V \sim 3.3$. The second episode is the luminosity function bump, where a mechanism of extra-mixing seems to be acting, decreasing further the Li abundance, a result not predicted by canonical stellar evolutionary models.

1.3.1 Lithium-Rich Giants

The low surface Li abundance in red giants that we expect from canonical stellar evolutionary models is what is observed for most of the giants. However, some of these stars have A(Li) > 1.5, a threshold that is used commonly to define giants that have gone through a noncanonical process of Li enrichment. Found for the first time by Wallerstein and Sneden (1982), we now can find several enriched giants in the field, clusters, dwarf galaxies, the galactic disc and bulge (e.g., da Silva et al. 1995;

Balachandran et al. 2000; Reddy and Lambert 2005; Gonzalez et al. 2009; Ruchti et al. 2011; Monaco et al. 2011; Lebzelter et al. 2012; Kirby et al. 2012; Martell and Shetrone 2013; Delgado Mena et al. 2016).

Li-rich giants, defined as those with A(Li) > 1.5, are a real population. They are not common, representing 1–2% of the giants, a percentage that depends on the exact sample and its location (e.g., Brown et al. 1989; Kumar et al. 2011). They are less frequent in globular clusters (Kirby et al. 2016; D'Orazi et al. 2015).

Some explanations for these unusual objects have been proposed, all based on noncanonical physics: either this is internal Li production in the star operating in the RGB, to which a very efficient extra-mixing mechanism needs to be added, or this is revealing interaction with external sources, such as planets, brown dwarfs, or binary companions.

Internal mechanisms require this efficient extra-mixing mechanism to rapidly transport the Li to a zone of the star with cooler temperatures, where it is not immediately burned by proton capture. In AGB stars, during the shell flashes, beryllium is produced in the convective envelope and thus it is convection what transports the Li to cooler regions. This process naturally happening in AGB stars is known as hot bottom burning (Sackmann and Boothroyd 1992). Instead, in the RGB phase (Sackmann and Boothroyd 1999), the shell where Li is created is not directly connected to the convective envelope, and an extra-mixing process needs to connect these two regions of the star so that we can observe the Li signature in the surface of stars. It is worth noticing that this very efficient extra-mixing mechanism needed to create Li-rich giants, so-called enhanced extra-mixing, is not the same as the canonical extra-mixing. Although the actual physical mechanism behind both processes is unknown and may be the same, their efficiencies need to be different, as the enhanced extra-mixing should be more efficient to produce the needed increase in the surface Li abundance of a very small percentage of giants, while the canonical extra-mixing affects all upper RGB giants and decreases their Li abundance.

The different possibilities for internal scenarios of Li enrichment are given by the different ways in which they transport Li or Be into the convective zone. Even if the canonical extra-mixing and the enhanced extra-mixing are not necessarily the same, both are expected to develop only after the RGB bump. This means that Li-rich giants cannot be created via internal production plus extra-mixing in the regions of parameter space where extra-mixing is less efficient, which is before the RGB bump and at higher metallicities.

The other proposed explanation for Li-rich giants are those where external sources share their material with the stars, and as such, we will refer to these as external mechanisms. The source of Li could be any contaminant that preserves or creates Li during their lifetimes, e.g., planets, brown dwarfs, or binary companions.

Other observational signatures and abundance patterns can be used to distinguish between Li replenishment scenarios. When engulfing a planet, the beryllium abundance of the star could also change. Melo et al. (2005) measure this element in a sample of stars with Li, finding no ^{9}Be. They argue that this is an indication of internal mechanisms acting, given that planet engulfment should increase both the Li and beryllium.

The other stable isotope of lithium, ^{6}Li, is far less abundant, but it has also been suggested that its absence in Li-rich giants favors the internal production plus extra-mixing mechanisms (e.g., Monaco et al. 2014; Drake et al. 2002). Given that ^{6}Li would be preserved in planets and brown dwarfs, the ingestion of these objects by the star would increase the ^{6}Li of the star. On the contrary, the internal scenarios only consider the production of the ^{7}Li isotope, and no change in the ^{6}Li should be expected in this case.

Even if the presence or absence of these two elements (^{6}Li and beryllium) in the star are often used in favor of one scenario of enrichment over the other, no models have really tested how the abundances of these elements should increase after the engulfment of a substellar mass companion.

The carbon isotopic ratio ($^{12}C/^{13}C$) is another tool to identify mechanisms. The FDU not only dilutes Li but also decreases the $^{12}C/^{13}C$ ratio. ^{13}C is accumulated closer to the burning shell, so extra-mixing reduces the carbon isotopic ratio, while engulfment will not impact it significantly. Carlberg et al. (2012) measured $^{12}C/^{13}C$ for a sample of stars consisting of their own measurements and stars from the Kumar et al. (2011) sample. Their stars have a carbon isotopic ratio consistent with planet accretion, but some stars in Kumar et al. (2011) could have experienced extra-mixing, as they have smaller $^{12}C/^{13}C$. Finding Li-rich stars with different carbon isotopic ratios could indicate that these unusual giants are created via different mechanisms, a combination of internal and external scenarios. Tayar et al. (2015) have measured the carbon isotopic ratio of a sample of rapidly rotating giants, finding high values that could be associated with winds from evolved giants or planet engulfment. It would be interesting to find the Li abundance of this sample.

Infrared (IR) excess is also unexpected in RGB stars (Zuckerman et al. 1995). de La Reza et al. (1996) proposed a scenario that linked the high Li abundance of red giants with the presence of shells or discs, where the mechanism producing the Li enhancement also produces mass loss. Thus, IR excess is another observational signature that has been associated with Li enrichment, but other authors (e.g., Bharat Kumar et al. 2015) have questioned this association. Rebull et al. (2015) find that if a red giant has a large IR excess, it is probably a Li-rich giant, but the correlation is not given the other way around.

The rotation could also help to distinguish between scenarios. Because of angular momentum conservation, once the radii of giants expand, they decrease their rotation rates. Thus, another type of unusual giants are those that rotate rapidly. Planet or brown dwarf engulfment can also increase the rotation rate of giants, and not also increase their Li abundances, because they are a good source of angular momentum (Carlberg et al. 2009). As planet engulfment can explain both enhancements in Li and rotation rates, a star with both anomalies could have been produced by this process. However, finding stars that are only enriched in Li but do not rotate rapidly is not enough to discard engulfment as the source of enrichment, because the rotation signal could fade when giants expand further in the RGB. Some works (Drake et al. 2002; Carlberg et al. 2012) have found correlations between the rapid rotation and Li enrichment in giants, where rapid rotators tend to be more enriched in Li.

Finally, to distinguish among scenarios of Li enrichment, we can use the evolutionary stage of the stars. This information is key. Initially, the scientific community focused on the idea that stars cluster around the RGB bump. With *Hipparcos* parallaxes, Charbonnel and Balachandran (2000) found that Li-rich giants appear on this region of the HR diagram, and as such, the destruction of the μ barrier could allow mixing to proceed and bring Li to the convective envelope. Based on this observational evidence, models also concentrated on explaining giants after the luminosity function bump (e.g., Denissenkov and Herwig 2004). If there is an association with the bump, it could not be explained by the planet engulfment mechanism that should not have a preferential region in the HR diagram to be produced. Nowadays, stars have been found all along the RGB and not necessarily close to the RGB bump.

The previous correlation between Li enrichment and evolutionary phase is not evident anymore (Gonzalez et al. 2009; Lebzelter et al. 2012), and several Li-rich giants have been found burning steadily helium in their cores (i.e., the horizontal branch, Kumar et al. 2011). Thus, there could also be another Li enrichment episode taking place closer to the end of the RGB phase or during the He-flash. An alternative scenario by Denissenkov (2012) indicates that these clump stars might be burning hydrogen in a shell but that due to extra-mixing they cross the HR diagram, getting to a position closer to the clump.

The evolutionary stage is an important information to find the Li enrichment mechanism. However, it is hard to obtain this information for the observed Li-rich giants that can be found in the HR diagram in the region where both clump giants and first ascending RGB stars are located closely together, so stars in these phases can be very easily confused. The future here is in the use of asteroseismology in Li-rich giants (Silva Aguirre et al. 2014; Jofré et al. 2015), which can give us information on the mass and evolutionary status at the same time.

Li-rich giants challenge the canonical theory of stellar evolution, and by understanding this anomaly, we could understand nonstandard processes acting in stars, modifying their evolution, such as planet interaction, extra-mixing, or the production of light elements in stellar interiors.

1.4 Outline of This Thesis

The lack of a clear explanation for the enriched giants and their connection with other observational signatures present an opportunity to study the noncanonical physics affecting the abundance of the stars, such as planet-star interaction and extra-mixing.

In this thesis, I will explore the different mechanisms of Li enrichment in giants. I attempt to constrain the mechanism that is acting on the red giants and identify specific objects or regions of parameter space where one specific noncanonical process acts.

In Part I of this thesis, we will focus on the theoretical aspect of the different scenarios of Li enrichment. Chapter 2 provides a compilation of extra-mixing mechanisms that could transport the produced Li from the H-burning shell to the convective envelope. We study the literature on this complex problem finding that none of the currently known mixing mechanisms are able to increase the surface Li abundance of the stars as much as it is needed and that they can only act, or are more efficient, after the luminosity function bump. As such, we conclude that current models are unable to provide a consistent explanation for all Li-rich giants.

Given that internal mechanisms seem to be insufficient, and that external mechanisms can work for a combination of stellar properties for which internal mechanisms fail, Chap. 3 focuses on the external mechanisms and the Li-rich giants produced by this scenario. Here, we discuss the possibility of binarity for the enriched giants found in the literature. However, known Li-rich giants with radial velocity measurements indicate that these stars do not have binary companions. Thus, a different source of Li must be considered, that could be planets or brown dwarfs. In Sect. 3.2, we model in detail the process of planet or brown dwarf engulfment, exploring a wide parameter space, including mass and initial composition of both planet and star, metallicity, and planetary orbital periods. By assuming the best possible conditions in order to preserve the lithium inside the star, we constrain the maximum enrichment that we can find in giants due to planet (or brown dwarf) engulfment. This limit of A(Li)$\approx$2.2, set an upper limit for Li-rich giants that could be formed by this specific mechanism.

Even more importantly, and a conclusion that can change the way in which we study this specific problem of lithium enrichment, I show that the mass is a relevant property in determining if a giant is enriched in Li. The traditional definition for Li-rich giants can be misleading, as it actually depends on the stellar mass. Massive stars without substellar mass companions could be considered anomalous because they have higher Li abundances regardless of the presence of planets. In contrast, giants engulfing companions could be flagged as normal even if they have engulfed a substellar mass companion. Additionally, we should be cautious when using the typical limit of A(Li) $\gtrsim$ 1.5 for enriched giants or any limit that does not take into consideration the mass. Assuming a higher limit in lithium abundance when selecting samples to study (e.g., assuming limits close to A(Li)$\approx$2.2 or a higher value), we are introducing a bias in our sample, selecting only giants that are not created by engulfment. This would lead to the possibly wrong conclusion that planet engulfment does not contribute significantly to the Li enrichment of giants. From here on, based on these conclusions, we should study this problem always considering the stellar masses, hopefully comparing the Li abundances of giant with stars of similar stellar parameters to find with precision which giants are enriched and study the problem correctly.

When modeling the Li enrichment of a giant due to the engulfment of planets or brown dwarfs, we also study the abundance of the less abundant isotope of lithium, ^{6}Li. We find that its absence should not be used as a tool to distinguish between enrichment scenarios for metal-rich stars, as ^{6}Li burns faster than ^{7}Li. Moreover, the absence of ^{6}Li in very ^{7}Li enriched stars does not give additional information about the engulfment scenario, as those stars could not be explained by substellar companion accretion only based on their ^{7}Li abundance. These theoretical conclusions are key to finding the mechanism behind the enrichment of Li in giants and will change how this problem is studied in the future.

Based on these models and related conclusions, I present interesting samples to study the Li enrichment mechanism in giants in Part II. The first of these consist of the giants in the open cluster Trumpler 20 (Chap. 4) that have measured Li abundances obtained by the *Gaia*-ESO survey. Two of the stars have anomalously high abundances, and I present an explanation, where Li is normal in the vast majority of cluster stars, except in these two cases, where the enrichment could be produced by the engulfment of a substellar mass companion. For the second interesting sample, I measured the Li abundance in 137 field giants with known atmospheric parameters in Chap. 5. This sample is interesting not only because of the measured mass for the giants but also because these stars are part of a planet search program, and as such, we know which of these giants currently have orbiting planets around them. Thus, we can link the enrichment mechanism with the presence of planets around the stars. With this sample, we show an example of how to find Li-enriched giants, by comparing with stars of similar properties, and find several possibly Li-enriched giant.

In a different approach to the problem of Li-enriched giants, and also as a possibility to study other major issues related to the Li abundance in stars, I present a sample of homogeneous Li abundances in dwarfs and subgiants in Chap. 6. This dataset allows, on one hand, to study the region known as the Lithium desert, a range in temperature where no stars are found with intermediate lithium abundances, and only high or low lithium are measured. We conclude here that the stars with low Li are slightly more evolved, being subgiants that have evolved from the Li dip. As such, these stars were already depleted previously and the absence of stars with intermediate Li in that range of temperature is expected and can be explained by simple stellar evolution. But, on the other hand, and relevant to the problem of Li enrichment in giants, these already Li-depleted subgiants that have evolved from the Li dip are ideal for testing the substellar companion engulfment scenario, making this sample a particularly interesting set of stars for future studies. A more detailed description of each of the samples can be found directly in the introduction in Part II.

Finally, a brief summary of the thesis, conclusions, and future work are presented in Part III.

References

M. Arnould, H. Norgaard, A&A **42**, 55 (1975)

M. Asplund, N. Grevesse, A.J. Sauval, P. Scott, ARA&A **47**, 481 (2009). http://doi.org/10.1146/annurev.astro.46.060407.145222

M. Asplund, D.L. Lambert, P.E. Nissen, F. Primas, V.V. Smith, ApJ **644**, 229 (2006). http://doi.org/10.1086/503538

S. Balachandran, ApJ **446**, 203 (1995). http://doi.org/10.1086/175779

S.C. Balachandran, F.C. Fekel, G.W. Henry, H. Uitenbroek, ApJ **542**, 978 (2000). http://doi.org/10.1086/317055

P. Baumann, I. Ramírez, J. Meléndez, M. Asplund, K. Lind, A&A **519**, A87 (2010). http://doi.org/10.1051/0004-6361/201015137
Y. Bharat Kumar, B.E. Reddy, C. Muthumariappan, G. Zhao, A&A **577**, A10 (2015). http://doi.org/10.1051/0004-6361/201425076
A.M. Boesgaard, K.G. Budge, M.E. Ramsay, ApJ **327**, 389 (1988). http://doi.org/10.1086/166201
A.M. Boesgaard, M.J. Tripicco, ApJL **302**, L49 (1986). http://doi.org/10.1086/184635
P. Bonifacio, P. Molaro, T. Sivarani, et al., A&A **462**, 851 (2007). http://doi.org/10.1051/0004-6361:20064834
J. Bouvier, A&A **489**, L53 (2008). http://doi.org/10.1051/0004-6361:200810574
R.N. Boyd, C.R. Brune, G.M. Fuller, C.J. Smith, PhRvD **82**(10), 105005 (2010). http://doi.org/10.1103/PhysRevD.82.105005
J.A. Brown, C. Sneden, D.L. Lambert, E. Dutchover, Jr., ApJS **71**, 293 (1989). http://doi.org/10.1086/191375
J.K. Carlberg, K. Cunha, V.V. Smith, S.R. Majewski, ApJ **757**, 109 (2012). http://doi.org/10.1088/0004-637X/757/2/109
J.K. Carlberg, S.R. Majewski, P. Arras, ApJ **700**, 832 (2009). http://doi.org/10.1088/0004-637X/700/1/832
M. Carlos, P.E. Nissen, J. Meléndez, A&A **587**, A100 (2016). http://doi.org/10.1051/0004-6361/201527478
M. Castro, S. Vauclair, O. Richard, N.C. Santos, A&A **494**, 663 (2009). http://doi.org/10.1051/0004-6361:20078928
J. Chanamé, M. Pinsonneault, D.M. Terndrup, ApJ **631**, 540 (2005). http://doi.org/10.1086/432410
C. Charbonnel, S.C. Balachandran, A&A **359**, 563 (2000)
C. Charbonnel, S. Talon, Science **309**, 2189 (2005). http://doi.org/10.1126/science.1116849
Y.Q. Chen, P.E. Nissen, T. Benoni, G. Zhao, A&A **371**, 943 (2001). http://doi.org/10.1051/0004-6361:20010371
J. Christensen-Dalsgaard, D.O. Gough, M.J. Thompson, ApJ **378**, 413 (1991). http://doi.org/10.1086/170441
D.D. Clayton, *Principles of stellar evolution and nucleosynthesis* (1968)
A. Coc, S. Goriely, Y. Xu, M. Saimpert, E. Vangioni, ApJ **744**, 158 (2012). http://doi.org/10.1088/0004-637X/744/2/158
A. Coc, J.P. Uzan, E. Vangioni, JCAP **10**, 050 (2014). http://doi.org/10.1088/1475-7516/2014/10/050
A. Coc, E. Vangioni, in *Journal of Physics Conference Series*, *Journal of Physics Conference Series*, vol. 202 (2010), *Journal of Physics Conference Series*, vol. 202, p. 012001. http://doi.org/10.1088/1742-6596/202/1/012001
R.H. Cyburt, M. Pospelov, Int. J. Modern Phys. E **21**, 1250004-1-1250004-13 (2012). http://doi.org/10.1142/S0218301312500048
L. da Silva, R. de La Reza, B. Barbuy, ApJL **448**, L41 (1995). http://doi.org/10.1086/309593
R. de La Reza, N.A. Drake, L. da Silva, ApJL **456**, L115 (1996). http://doi.org/10.1086/309874
E. Delgado Mena, G. Israelian, J.I. González Hernández, et al., A&A **562**, A92 (2014). http://doi.org/10.1051/0004-6361/201321493
E. Delgado Mena, S. Bertrán de Lis, V.Z. Adibekyan, et al., A&A **576**, A69 (2015). http://doi.org/10.1051/0004-6361/201425433
E. Delgado Mena, M. Tsantaki, S.G. Sousa, et al., A&A **587**, A66 (2016). http://doi.org/10.1051/0004-6361/201527196
P.A. Denissenkov, ApJL **753**, L3 (2012). http://doi.org/10.1088/2041-8205/753/1/L3
P.A. Denissenkov, F. Herwig, ApJ **612**, 1081 (2004). http://doi.org/10.1086/422575
P.A. Denissenkov, D.A. VandenBerg, ApJ **593**, 509 (2003). http://doi.org/10.1086/376410
V. D'Orazi, R.G. Gratton, G.C. Angelou, et al., ApJL **801**, L32 (2015). http://doi.org/10.1088/2041-8205/801/2/L32
N.A. Drake, R. de la Reza, L. da Silva, D.L. Lambert, AJ **123**, 2703 (2002). http://doi.org/10.1086/339968

P. Eggenberger, A. Maeder, G. Meynet, A&A **519**, L2 (2010). http://doi.org/10.1051/0004-6361/201014939
B.D. Fields, Annu. Rev. Nucl. Part. Sci. **61**, 47 (2011). http://doi.org/10.1146/annurev-nucl-102010-130445
P. Figueira, J.P. Faria, E. Delgado-Mena, et al., A&A **570**, A21 (2014). http://doi.org/10.1051/0004-6361/201424218
X. Fu, A. Bressan, P. Molaro, P. Marigo, MNRAS **452**, 3256 (2015). http://doi.org/10.1093/mnras/stv1384
L. Ghezzi, K. Cunha, V.V. Smith, R. de la Reza, ApJ **724**, 154 (2010). http://doi.org/10.1088/0004-637X/724/1/154
G. Gonzalez, MNRAS **386**, 928 (2008). http://doi.org/10.1111/j.1365-2966.2008.13067.x
G. Gonzalez, MNRAS **446**, 1020 (2015). http://doi.org/10.1093/mnras/stu2156
O.A. Gonzalez, M. Zoccali, L. Monaco, et al., A&A **508**, 289 (2009). http://doi.org/10.1051/0004-6361/200912469
R.G. Gratton, C. Sneden, E. Carretta, A. Bragaglia, A&A **354**, 169 (2000)
G. Guiglion, P. de Laverny, A. Recio-Blanco, et al., ArXiv e-prints: 1608.03411 (2016)
L.M. Hobbs, C. Pilachowski, ApJL **309**, L17 (1986). http://doi.org/10.1086/184752
I. Iben, Jr., ApJ **147**, 624 (1967). http://doi.org/10.1086/149040
G. Israelian, N.C. Santos, M. Mayor, R. Rebolo, A&A **414**, 601 (2004). http://doi.org/10.1051/0004-6361:20034398
G. Israelian, E. Delgado Mena, N.C. Santos, et al., Nature **462**, 189 (2009). http://doi.org/10.1038/nature08483
E. Jofré, R. Petrucci, L. García, M. Gómez, A&A **584**, L3 (2015). http://doi.org/10.1051/0004-6361/201527337
B.F. Jones, D. Fischer, M. Shetrone, D.R. Soderblom, AJ **114**, 352 (1997). http://doi.org/10.1086/118479
R. Kippenhahn, A. Weigert, *Stellar Structure and Evolution* (1990)
R. Kippenhahn, A. Weigert, E. Hofmeister, Methods Comput. Phys. **7**, 129 (1967)
E.N. Kirby, X. Fu, P. Guhathakurta, L. Deng, ApJL **752**, L16 (2012). http://doi.org/10.1088/2041-8205/752/1/L16
E.N. Kirby, P. Guhathakurta, A.J. Zhang, et al., ApJ **819**, 135 (2016). http://doi.org/10.3847/0004-637X/819/2/135
E. Komatsu, K.M. Smith, J. Dunkley, et al., ApJS **192**, 18 (2011). http://doi.org/10.1088/0067-0049/192/2/18
Y.B. Kumar, B.E. Reddy, D.L. Lambert, ApJL **730**, L12 (2011). http://doi.org/10.1088/2041-8205/730/1/L12
D.L. Lambert, B.E. Reddy, MNRAS **349**, 757 (2004). http://doi.org/10.1111/j.1365-2966.2004.07557.x
T. Lebzelter, S. Uttenthaler, M. Busso, M. Schultheis, B. Aringer, A&A **538**, A36 (2012). http://doi.org/10.1051/0004-6361/201117743
K. Lind, F. Primas, C. Charbonnel, F. Grundahl, M. Asplund, A&A **503**, 545 (2009). http://doi.org/10.1051/0004-6361/200912524
K. Lodders, H. Palme, H.P. Gail, Landolt Börnstein (2009). http://doi.org/10.1007/978-3-540-88055-4_34
R.E. Luck, U. Heiter, AJ **131**, 3069 (2006). http://doi.org/10.1086/504080
S.L. Martell, M.D. Shetrone, MNRAS **430**, 611 (2013). http://doi.org/10.1093/mnras/sts661
S.L. Martell, G.H. Smith, M.M. Briley, AJ **136**, 2522 (2008). http://doi.org/10.1088/0004-6256/136/6/2522
C.H.F. Melo, P. de Laverny, N.C. Santos, et al., A&A **439**, 227 (2005). http://doi.org/10.1051/0004-6361:20041805
G. Michaud, ApJ **302**, 650 (1986). http://doi.org/10.1086/164025
L. Monaco, S. Villanova, C. Moni Bidin, et al., A&A **529**, A90 (2011). http://doi.org/10.1051/0004-6361/201016285

L. Monaco, H.M.J. Boffin, P. Bonifacio, et al., A&A **564**, L6 (2014). http://doi.org/10.1051/0004-6361/201323348
J. Montalbán, E. Schatzman, A&A **354**, 943 (2000)
S. Palmerini, S. Cristallo, M. Busso, et al., ApJ **741**, 26 (2011). http://doi.org/10.1088/0004-637X/741/1/26
M.H. Pinsonneault, C.P. Deliyannis, P. Demarque, ApJS **78**, 179 (1992). http://doi.org/10.1086/191624
Planck Collaboration, P.A.R. Ade, N. Aghanim, et al., A&A **594**, A13 (2016). http://doi.org/10.1051/0004-6361/201525830
M. Pospelov, J. Pradler, Annu. Rev. Nucl. Part. Sci. **60**, 539 (2010). http://doi.org/10.1146/annurev.nucl.012809.104521
N. Prantzos, A&A **542**, A67 (2012). http://doi.org/10.1051/0004-6361/201219043
I. Ramírez, J.R. Fish, D.L. Lambert, C. Allende Prieto, ApJ **756**, 46 (2012). http://doi.org/10.1088/0004-637X/756/1/46
S. Randich, R. Gratton, R. Pallavicini, L. Pasquini, E. Carretta, A&A **348**, 487 (1999)
L.M. Rebull, J.K. Carlberg, J.C. Gibbs, et al., AJ **150**, 123 (2015). http://doi.org/10.1088/0004-6256/150/4/123
B.E. Reddy, D.L. Lambert, AJ **129**, 2831 (2005). http://doi.org/10.1086/430190
H. Reeves, Nature **226**, 727 (1970). http://doi.org/10.1038/226727a0
D. Romano, F. Matteucci, P. Ventura, F. D'Antona, A&A **374**, 646 (2001). http://doi.org/10.1051/0004-6361:20010751
G.R. Ruchti, J.P. Fulbright, R.F.G. Wyse, et al., ApJ **743**, 107 (2011). http://doi.org/10.1088/0004-637X/743/2/107
S.G. Ryan, MNRAS **316**, L35 (2000). http://doi.org/10.1046/j.1365-8711.2000.03777.x
S.G. Ryan, T.C. Beers, C.P. Deliyannis, J.A. Thorburn, ApJ **458**, 543 (1996). http://doi.org/10.1086/176838
I.J. Sackmann, A.I. Boothroyd, ApJL **392**, L71 (1992). http://doi.org/10.1086/186428
I.J. Sackmann, A.I. Boothroyd, ApJ **510**, 217 (1999). http://doi.org/10.1086/306545
L. Sbordone, P. Bonifacio, E. Caffau, et al., A&A **522**, A26 (2010). http://doi.org/10.1051/0004-6361/200913282
A. Serenelli, Eur. Phys. J. A **52**, 78 (2016). http://doi.org/10.1140/epja/i2016-16078-1
M.D. Shetrone, ApJL **585**, L45 (2003). http://doi.org/10.1086/374262
V. Silva Aguirre, G.R. Ruchti, S. Hekker, et al., ApJL **784**, L16 (2014). http://doi.org/10.1088/2041-8205/784/1/L16
F. Spite, M. Spite, A&A **115**, 357 (1982)
M. Spite, R. Cayrel, B. Plez, et al., A&A **430**, 655 (2005). http://doi.org/10.1051/0004-6361:20041274
F.J. Swenson, J. Faulkner, ApJ **395**, 654 (1992). http://doi.org/10.1086/171686
A. Tajitsu, K. Sadakane, H. Naito, A. Arai, W. Aoki, Nature **518**, 381 (2015). http://doi.org/10.1038/nature14161
J.L. Tassoul, *Stellar Rotation* (2007)
J. Tayar, T. Ceillier, D.A. García-Hernández, et al., ApJ **807**, 82 (2015). http://doi.org/10.1088/0004-637X/807/1/82
C. Travaglio, S. Randich, D. Galli, et al., ApJ **559**, 909 (2001). http://doi.org/10.1086/322415
R.V. Wagoner, W.A. Fowler, F. Hoyle, ApJ **148**, 3 (1967). http://doi.org/10.1086/149126
G. Wallerstein, C. Sneden, ApJ **255**, 577 (1982). http://doi.org/10.1086/159859
S.E. Woosley, D.H. Hartmann, R.D. Hoffman, W.C. Haxton, ApJ **356**, 272 (1990). http://doi.org/10.1086/168839
D.R. Xiong, L. Deng, MNRAS **395**, 2013 (2009). http://doi.org/10.1111/j.1365-2966.2009.14581.x
B. Zuckerman, S.S. Kim, T. Liu, ApJL **446**, L79 (1995). http://doi.org/10.1086/187935

Part I
Scenarios of Lithium Enrichment

Chapter 2
Internal Mechanisms

Li production through the Cameron-Fowler mechanism (Cameron and Fowler 1971) plus extra-mixing is one of the possible mechanisms that can increase the Li abundance in the outer convective envelope of giants.

During the Cameron-Fowler nuclear chain, the ^{3}He inside the star produces beryllium, which is then converted into ^{7}Li, after the beryllium decays by electron capture:

$$^3\text{He} + {}^4\text{He} \rightarrow {}^7\text{Be} + \gamma, \tag{2.1}$$

$$^7\text{Be} + \text{e}^- \rightarrow {}^7\text{Li} + \nu. \tag{2.2}$$

These reactions naturally occur inside main sequence stars during the hydrogen fusion to create ^{4}He, where, due to the high temperatures, ^{7}Li gets rapidly destroyed by capturing protons. The survival of the ^{7}Li created inside the star depends on how quickly the element or its parent beryllium can be transported outward, toward zones of the star with cooler temperatures, where Li can survive.

As was briefly introduced in Sect. 1.3.1, the internal mechanisms to produce Li enrichment all depend on the Li production by Cameron-Fowler mechanism but differ in how the Li is transported to the convective layer, a process known as "enhanced extra-mixing" to distinguish it from the canonical extra-mixing that decreases the Li abundance and changes the C, N, O, and carbon isotopic ratios after the luminosity function bump.

As with the canonical extra-mixing, several possibilities have been suggested for this "enhanced extra-mixing" that could contribute to explain Li-rich giants. We will briefly explain each of these possibilities and summarize current models found in the literature. Then, we can conclude if any of the mixing mechanisms by themselves could explain the enriched giants observed. In general, the physical mechanisms behind the mixing attributed to the enhanced extra-mixing are the same suggested for the canonical extra-mixing. However, even if both mixing events are physically

C. Aguilera-Gómez, *Explaining Lithium Enriched Red Giant Branch Stars*,
Springer Theses, https://doi.org/10.1007/978-3-030-02583-0_2

motivated by the same process, the efficiency of the mixing must be much larger in the enhanced extra-mixing to increase the surface Li abundance of stars.

The first theoretical approach to Li-rich giants produced by extra-mixing is the parametrized conveyor belt models of Sackmann and Boothroyd (1999). Here, they model the mixing as a deep circulation process with free parameters, some of which are fitted based on observations of $^{12}C/^{13}C$ in RGB stars. Basically, matter from the bottom of the convective layer travels downward in the star reaching higher temperatures and later returning upward, mixing with the envelope. They consider that the mixing starts after the RGB bump and continues until the star reaches the RGB tip and find that, under very specific conditions, a large amount of lithium can be produced during the RGB phase, reaching values of $A(Li) = 4.0$. To produce Li instead of destroying it, very rapid circulation is required and the final abundance is highly dependent on geometry as well. In this work, the authors do not suggest any physical mechanism that could work based on the needed parameters obtained by their modeling. Although these models do not consider a physical mechanism behind the mixing, they do suggest that it is possible to increase the surface Li abundance. Since then, several possible mixing mechanisms have been suggested as candidates.

2.1 Rotationally Induced Mixing

It is known that rotation can induce mixing inside stars; thus, it has been considered as a possible physical driving mechanism for the canonical and enhanced extra-mixing.

Meridional circulation, consisting of large-scale currents that rise at the pole and descend at the equator, is produced because the star cannot maintain hydrostatic equilibrium in the radiative zones when it is rotating (Eddington 1925; Zahn 1992), as the gravity is smaller at the equator than at the poles. These currents transport chemical elements, thus mixing the interior of the star (Sweigart and Mengel 1979). Other types of instabilities can also be produced inside a rotating star and drive mixing, such as shear instabilities that develop in a differentially rotating star (Zahn 1974).

Extra-mixing by rotation in these two different forms has been considered extensively as the cause of changes in the surface composition of RGB stars beyond the bump (Charbonnel 1995; Denissenkov and Tout 2000; Denissenkov and VandenBerg 2003). The diffusion coefficient (and efficiency) that characterizes this mixing increases with rotation rate and also depends on the degree of differential rotation inside the star.

Regarding the creation of Li-rich giants, Palacios et al. (2001) suggested the idea of a "Li-flash." In this process, ^3He is converted into ^7Be in a shell with high enough temperature. Then, once this shell crosses the molecular weight gradient (and the star is located in the luminosity function bump), the beryllium is transported to cooler regions of the star by the canonical extra-mixing, where Li can be produced. Then, this recently created Li is transported to hotter layers of the star by meridional circulation, where it will burn by proton capture. They claim that the energy generation rate of this reaction is higher than that of the H-burning shell because of its efficiency. Then, this new fresh source of energy increases greatly the velocity of the extra-mixing, to the extent necessary to produce an increase of Li in the surface of the star. According to this hypothesis, all giants should go through a short-lived enrichment phase. The Li production through the Li-flash could not be confirmed by the models of Denissenkov and Herwig (2004). They find that extra-mixing does cause a second burning shell slightly above the H-burning shell, but the main source of energy there is the burning of ^{3}He and not Li. And the presence of this shell does not affect the surface Li abundance, because the diffusion coefficient only increases in a narrow mass interval, far from where Li is produced. As such, they report that canonical extra-mixing cannot produce an increase in Li abundance. Based on the large diffusion coefficient that was needed to produce Li enrichment, according to previous modeling (Denissenkov and Weiss 2000), they can obtain short-lived increases of Li abundance. This is a parametric approach, but they do consider a possible physical mechanism behind this behavior. If the extra-mixing in RGB stars is triggered by rotation, the Li production can be triggered by an increase in rotation rate by an external source.

Although all these works considered the possibility of canonical and enhanced extra-mixing induced by rotation, the latest results indicate that meridional circulation and shear instabilities do not get high enough mixing rates to reproduce the anomalous observed abundances or increase the Li in the envelope of stars. Chanamé et al. (2005) follow the rotational evolution of low-mass RGB stars with a fully self-consistent model. They find that regardless of the presence of the mean molecular weight barrier, rotationally induced mixing exists but it is extremely inefficient on the lower RGB; thus, in contrast with previous works, they do not require the erasing of the μ-barrier to drive mixing. An interesting conclusion in this work is that, to account for observations, the diffusion coefficients (i.e., the strength of the mixing) need to be much larger than what previous theoretical computations had considered but also that rapidly rotating giants need to have a lower efficiency of rotational mixing, which could be accomplished via the interaction between convection in the envelope of the giants and rotation.

With this work (Chanamé et al. 2005), it already seems as rotational-mixing is inefficient even when considering only canonical extra-mixing. Palacios et al. (2006) confirm this with their models. They used a model that considers the transport of angular momentum by shear instabilities and meridional circulation, and, as was previously suggested, a larger mixing strength is found in a differentially rotating convective zone. But even then, when mixing is maximized, it is not enough to explain the chemical abundance pattern in RGB stars. It is important to notice that this conclusion invalidates the Li-flash scenario, as they never obtained the mixing rates needed to start the process.

The left panel of Fig. 2.1, obtained from Palacios et al. (2006), compares observations of surface chemical abundances of different elements (asterisk as actual measurements, open triangles as lower limits, and open downward triangles as upper limits) with what is obtained by their models of rotationally induced mixing (different lines for different prescriptions of rotation). These include the Li

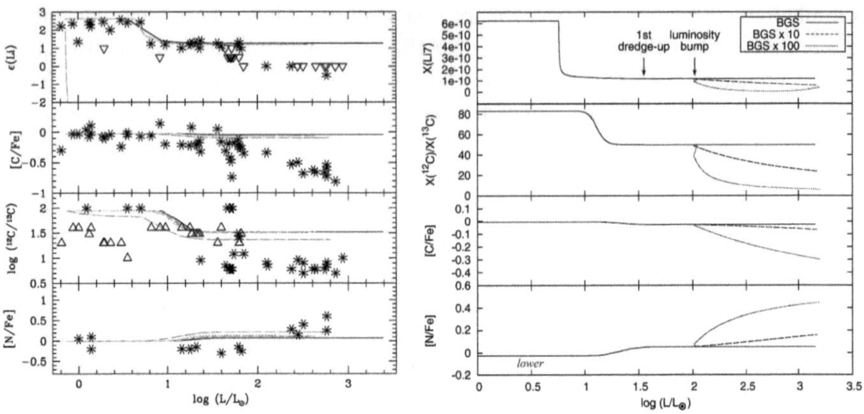

Fig. 2.1 Left panel: Figure 15 from Palacios et al. (2006). Surface abundances of Li, [C/Fe], carbon isotopic ratio, and [N/Fe] for models (lines) that include a consistent treatment of rotation. These are compared with observations from Gratton et al. (2000), where asterisks are measurements, triangles are lower limits, and downward triangles are upper limits. Mixing by rotation does not predict changes in abundances after the luminosity bump and cannot explain the chemical abundance pattern observed. Credit: Palacios et al. 2006, A&A, 453, 261, reproduced with permission ©ESO. Right panel: Figure 4 from Wachlin et al. (2014). Surface abundances of Li, carbon isotopic ratio, [C/Fe], and [N/Fe] for models with thermohaline mixing (BGS) using the prescription of Brown et al. (2013), and cases where the diffusion coefficient was artificially increased by factors of 10 (dashed line) or 100 (dotted line). Credit: Wachlin et al. 2014, A&A, 570, A58, reproduced with permission ©ESO

abundance in the top panel (where $\epsilon(Li)$ is another notation for A(Li)), showing the typical decrease after the bump expected to trigger due to canonical extra-mixing. These models, which include a consistent treatment of rotation, agree with data until the luminosity function bump. For higher luminosities, the diffusion coefficients are too small to have an effect on the surface composition and are not able to reproduce the chemical abundance pattern observed. This figure does not directly depict the relation of rotation with Li-rich giants, but it is known that, to produce an enrichment, larger diffusion coefficients are needed. As the coefficients are not even large enough to explain the decrease in Li, this invalidates rotation as an origin for the enrichment.

In conclusion, it seems that mixing by meridional circulation and shear instabilities is unable to explain the buildup of Li in stellar envelopes.

2.2 Thermohaline Mixing

Another potential mixing mechanism acting to change the surface abundance is thermohaline mixing (or fingering convection) where the reaction

$$^{3}\mathrm{He} + {}^{3}\mathrm{He} \rightarrow {}^{4}\mathrm{He} + 2\mathrm{p} \quad (2.3)$$

creates an inversion of molecular weight gradient with depth, triggering an instability analogous to salt-finger ones in the ocean.

This process has been studied in oceanography for a long time, but thermohaline convection has just recently been considered as a viable mixing mechanism in stars. Eggleton et al. (2006) found an inversion of the chemical composition gradient in their stellar model after the luminosity function bump, which led Charbonnel and Zahn (2007) to conclude that the process of thermohaline mixing can significantly change the surface composition of RGB stars.

Thermohaline mixing was thought to be a good candidate for the canonical extra-mixing that decreases the Li abundance, since this process develops only after the luminosity bump and it predicts similar changes in surface composition of stars of similar masses and metallicities (because the size of the inversion in μ depends on the amount of ^{3}He only), something that has been observed in RGB stars, although it is not well understood how it can change the surface abundance of Li and other elements like carbon and nitrogen.

The difficulty with thermohaline mixing so far has been to model the exact conditions found inside stars in 2D and in 3D. Assuming that matter moves in relatively narrow fingers and in a vertical direction, Ulrich (1972) developed a linear theory, modeling the mixing with a diffusion coefficient that depends on a parameter for the efficiency of the mixing, which, at the same time, is strongly dependent on the geometry of the fingers. This linear theory has been used by several works to treat the thermohaline mixing. Charbonnel and Zahn (2007) use it and indicate that models need very elongated fingers to account for observations of chemical patterns in the RGB. As laboratory experiments that replicate conditions found in the ocean show (Krishnamurti 2003), this seems to be the case for the fingers formed in the ocean, but conditions inside an RGB star can be significantly different. The lack of a good estimation for the aspect ratio of fingers affects the diffusion coefficient. Thus, an overestimated finger size would increase the mixing efficiency and produce models that can explain observations but may be considering unrealistic physics inside stars.

Denissenkov (2010) presents 2D numerical simulations of thermohaline mixing and discusses the possible creation of Li-rich giants by this mechanism. In this case, a very high value of the diffusion coefficient would be needed to transport the Li to the convective layer and increase the abundance (instead of transporting effectively the Li from the surface to the interior and decreasing the surface abundance). To reach such high values of diffusion coefficient, an unrealistic aspect ratio of the fingers would be needed. The author also mentions that, although it would seem that thermohaline convection would need to be replaced by some other mixing mechanism in Li-rich giants, it would be hard to explain why this other process would only affect Li-rich giants and not operate universally in all other upper RGB stars.

According to 1D models (e.g. Charbonnel and Zahn 2007; Denissenkov 2010; Angelou et al. 2011), thermohaline mixing could reproduce observations in upper RGB stars (being the physical explanation behind canonical extra-mixing), but 2D simulations (Denissenkov 2010) and new prescriptions (see Wachlin et al. 2011) clearly indicate otherwise. That being said, for Li-rich giants, a much bigger aspect ratio of fingers is needed. Thus, the difficulty to explain Li production in these stars is even larger based on the latest results.

To distinguish which aspect ratio of fingers correspond to those formed under the physical conditions found in stars, 3D numerical simulations may provide a solution. Denissenkov and Merryfield (2011) present three-dimensional computations of the process and confirm results of 2D simulations, as do 3D numerical hydrodynamical simulations approaching stellar conditions by Traxler et al. (2011). Although new theoretical models near the astrophysical regime (Brown et al. 2013) have found that previous simulations were underresolved, they confirm the results, indicating that thermohaline mixing is not responsible for the canonical extra-mixing, and consequently, it is not responsible for the enhanced extra-mixing either, which needs to be much more efficient. Worried about the lack of convergence on the latest results of thermohaline mixing theory, Wachlin et al. (2014) perform new simulations with a new prescription and under stellar conditions. Although the mixing coefficient obtained here is larger, and the fingers reach the convective zone, the efficiency of the mixing is still too low to account for observations of RGB stars.

Modeling thermohaline mixing is not an easy task even in 1D, especially when dealing with Li abundances. Lattanzio et al. (2015) emphasize how the results can drastically change depending on the numerical details of the models, regarding the production or destruction of Li in RGB stars. Here they find that Li can be produced inside the star with a much smaller (but still considerable) aspect ratio than the one predicted from 2D or 3D simulations and it is mostly created near the tip of the RGB. But the most important conclusion from this work is that the Li abundance obtained in models depends on numerical conditions and must be treated with caution.

As it appears that thermohaline mixing by itself struggles to explain canonical extra-mixing (and even more to explain Li-rich giants), some authors have considered models where the mixing is not only driven by thermohaline mixing, but there are also other mixing mechanisms included, like rotationally induced mixing or mixing driven by magnetic fields. Cantiello and Langer (2010) find that the effect of both of these mechanisms is negligible when compared to fingering convection, but they are not able to consider the interaction between mechanisms. Charbonnel and Lagarde (2010) add the mixing coefficients of thermohaline and rotationally induced mixing to consider both mechanisms and obtain results that agree with observations of RGB stars. This simplified treatment may not be entirely accurate, as the mixing coefficient needs to be completely recalculated when considering rotation and fingering convection, since horizontal turbulence would reduce the effect of thermohaline mixing (Denissenkov and Pinsonneault 2008; Vauclair and Théado 2012). Denissenkov (2010) indicates that adding the effect of any other mixing mechanisms would stratify the radiative zones and prevent the diffusive instability of thermohaline convection to develop.

The chemical abundance pattern produced by thermohaline mixing, modeled by Wachlin et al. (2014), can be seen in the right panel of Fig. 2.1. The solid line is obtained by using one of the latest 1D treatments of fingering convection that is based on 3D simulations (Brown et al. 2013). We see that it almost does not change the abundance after the luminosity function bump since its efficiency is too low. For comparison, they show results with diffusion coefficients artificially incremented by 10 or 100, which would be needed to explain the chemical abundance patterns observed, e.g., the decrease of Li mass fraction after the luminosity function bump (top panel). As we have explained throughout this chapter, the mixing needed to explain Li-rich giants must be even faster than that needed to explain the depletion after the bump. If diffusion coefficients are not high enough to account for canonical extra-mixing, they cannot explain enhanced extra-mixing either.

This once promising mixing mechanism also is insufficient to explain high Li abundance in RGB stars.

2.3 Magnetic Mechanisms and Magneto-Thermohaline Mixing

Busso et al. (2007) analyze if buoyant magnetic flux tubes, arising near the zone of hydrogen burning, can be a possible mixing mechanism. For this process to take place, a substantial toroidal magnetic field is needed. They found that this is a plausible scenario to decrease the surface Li abundance of RGB stars, under the assumption that the star has a large magnetic field. Magnetohydrodynamic modeling would be an interesting way to test this model.

The lack of explanation for Li-rich giants is one of the motivations for the development of an alternative mixing mechanism, magneto-thermohaline mixing (Denissenkov et al. 2009). The mechanism would need to be from 10 to 100 times faster than canonical extra-mixing to be efficient enough to produce Li enrichment. And, since rotationally induced mixing would suppress the effect of thermohaline mixing, other options are needed. Moreover, considering works that found that Li-rich giants tend to be faster rotators (Drake et al. 2002; Carlberg et al. 2012), then it's even harder to explain Li-rich giants with thermohaline mixing and rotationally induced mixing.

Denissenkov et al. (2009) use a model of magnetic buoyancy, where a toroidal magnetic field is generated due to the differentially rotating radiative zone of the giant. This field forms magnetic flux rings that can rise up to the convective envelope and produce mixing. In this work, in contrast with pure magnetic buoyancy models (Busso et al. 2007), they consider that the rings are formed in a region where μ is lower due to the burning of ^{3}He (^{3}He(^{3}He, ^{4}He)2p, the same reaction that would trigger thermohaline mixing), in which case their buoyancy is weakly dependent on the strength of the magnetic field and is controlled by the difference in mean molecular weight between the rings and the surrounding medium instead of the deficit in density caused by the extra magnetic pressure.

This new approach has a specific advantage over pure thermohaline mixing when referring to the creation of Li-rich giants, or, more specifically, fastly rotating Li-rich giants. The growth of fingers in the thermohaline mixing model is dampened by fast rotation, but magnetic flux rings are supposed to form more efficiently with rapid rotation.

An attempt to explain Li-rich giants with magnetic buoyancy was made by Guandalini et al. (2009). They assume two models of magnetic buoyancy: the first, where the magnetic field is growing and only local instabilities are produced near the H-burning shell, and the second, where the fields are large enough that the entire flux rings become buoyant. In this second stage, the speed of the rings is reduced due to heat exchange with the surrounding medium. As there is not enough information about the behavior of the RGB magnetic fields, they assume that in the bump they can apply the model of a growing magnetic field and then switch to the second mode. They obtain that the Li can increase in the first phase up to A(Li) =2.0–2.5, a value set by the equilibrium beryllium abundance near the H-burning shell. In general, the production or destruction of Li in these models still depends on several unknown parameters, as the velocity of the mixing, its maximum depth, and the time it has to develop.

The first attempts on magneto-thermohaline mixing and magnetic buoyancy look very promising but need further verification, hopefully in full 3D, but even then, if the maximum Li abundance that can be produced using this mixing mechanism is A(Li) $= 2.5$ as suggested by previous studies, something else is needed to explain some of the Li-rich giants, especially those with super-meteoritic abundances.

2.4 Summary

Internal mechanisms of Li production combine the production of the element via the Cameron-Fowler mechanism plus an efficient mixing process that transports the material from the H-burning shell to the convective envelope.

Although the physical driving force behind this enhanced extra-mixing might be the same as for canonical extra-mixing, it is clear that different strengths of the mixing are required to either increase or decrease the Li. Rotationally induced mixing, thermohaline mixing, and magnetic buoyancy have been extensively studied in the literature as possibilities, but no physical mechanism of mixing has been identified so far that is efficient enough to increase the Li abundance and reach the values observed in Li-rich giants, which can even have super-meteoritic abundances (A(Li) > 3.3).

Even if a mixing mechanism is ever found that can explain the high Li abundances in giants by itself, it should also be considered as a way to decrease the Li abundance and be a candidate for the canonical extra-mixing, as there is no reason why a mixing mechanism could operate only in a small subset of giants. At the same time, any mixing mechanism that is invoked to explain Li-rich giants needs to specify why the efficiency of the mixing is so high in these and not in all other giants.

We will see in Chap. 3 that external mechanisms by themselves are not able to explain the most enriched giants either. Thus, a possibility that must be considered is a combination of both the internal and external mechanisms. Denissenkov and Herwig (2004) consider that the accretion of external material can increase the rotation rates that enhance the extra-mixing and thus could explain the high Li abundances. It is also possible, although unlikely, that the engulfment of a companion enriched in ^{3}He could increase the efficiency of thermohaline mixing while injecting angular momentum that could explain the fast rotation and the possible correlation between rapid rotation and Li enhancement, as suggested by Denissenkov (2010), who however considered this scenario too complicated and highly unlikely. The interplay between mixing mechanisms and accretion of external material to produce Li-rich giants should be considered in more detail.

We emphasize that the extra-mixing mechanisms are expected to develop efficiently after the RGB bump. Thus, there is a domain in stellar parameters, where the internal mechanisms of Li production struggle to create Li-rich giants, and some other mechanism should be considered.

References

G.C. Angelou, R.P. Church, R.J. Stancliffe, J.C. Lattanzio, G.H. Smith, ApJ **728**, 79 (2011). http://doi.org/10.1088/0004-637X/728/2/79

J.M. Brown, P. Garaud, S. Stellmach, ApJ **768**, 34 (2013). http://doi.org/10.1088/0004-637X/768/1/34

M. Busso, G.J. Wasserburg, K.M. Nollett, A. Calandra, ApJ **671**, 802 (2007). http://doi.org/10.1086/522616

A.G.W. Cameron, W.A. Fowler, ApJ **164**, 111 (1971). http://doi.org/10.1086/150821

M. Cantiello, N. Langer, A&A **521**, A9 (2010). http://doi.org/10.1051/0004-6361/201014305

J.K. Carlberg, K. Cunha, V.V. Smith, S.R. Majewski, ApJ **757**, 109 (2012). http://doi.org/10.1088/0004-637X/757/2/109

J. Chanamé, M. Pinsonneault, D.M. Terndrup, ApJ **631**, 540 (2005). http://doi.org/10.1086/432410

C. Charbonnel, ApJL **453**, L41 (1995). http://doi.org/10.1086/309744

C. Charbonnel, N. Lagarde, A&A **522**, A10 (2010). http://doi.org/10.1051/0004-6361/201014432

C. Charbonnel, J.P. Zahn, A&A **467**, L15 (2007). http://doi.org/10.1051/0004-6361:20077274

P.A. Denissenkov, ApJ **723**, 563 (2010). http://doi.org/10.1088/0004-637X/723/1/563

P.A. Denissenkov, F. Herwig, ApJ **612**, 1081 (2004). http://doi.org/10.1086/422575

P.A. Denissenkov, W.J. Merryfield, ApJL **727**, L8 (2011). http://doi.org/10.1088/2041-8205/727/1/L8

P.A. Denissenkov, M. Pinsonneault, ApJ **684**, 626-634 (2008). http://doi.org/10.1086/590102

P.A. Denissenkov, M. Pinsonneault, K.B. MacGregor, ApJ **696**, 1823 (2009). http://doi.org/10.1088/0004-637X/696/2/1823

P.A. Denissenkov, C.A. Tout, MNRAS **316**, 395 (2000). http://doi.org/10.1046/j.1365-8711.2000.03498.x

P.A. Denissenkov, D.A. VandenBerg, ApJ **593**, 509 (2003). http://doi.org/10.1086/376410

P.A. Denissenkov, A. Weiss, A&A **358**, L49 (2000)

N.A. Drake, R. de la Reza, L. da Silva, D.L. Lambert, AJ **123**, 2703 (2002). http://doi.org/10.1086/339968

A.S. Eddington, The Observatory **48**, 73 (1925)

P.P. Eggleton, D.S.P. Dearborn, J.C. Lattanzio, Science **314**, 1580 (2006). http://doi.org/10.1126/science.1133065

R.G. Gratton, C. Sneden, E. Carretta, A. Bragaglia, A&A **354**, 169 (2000)

R. Guandalini, S. Palmerini, M. Busso, S. Uttenthaler, PASA **26**, 168 (2009). http://doi.org/10.1071/AS08063

R. Krishnamurti, J. Fluid Mech. **483**, 287 (2003). http://doi.org/10.1017/S0022112003004166
J.C. Lattanzio, L. Siess, R.P. Church, et al., MNRAS **446**, 2673 (2015). http://doi.org/10.1093/mnras/stu2238
A. Palacios, C. Charbonnel, M. Forestini, A&A **375**, L9 (2001). http://doi.org/10.1051/0004-6361:20010903
A. Palacios, C. Charbonnel, S. Talon, L. Siess, A&A **453**, 261 (2006). http://doi.org/10.1051/0004-6361:20053065
I.J. Sackmann, A.I. Boothroyd, ApJ **510**, 217 (1999). http://doi.org/10.1086/306545
A.V. Sweigart, J.G. Mengel, ApJ **229**, 624 (1979). http://doi.org/10.1086/156996
A. Traxler, P. Garaud, S. Stellmach, ApJL **728**, L29 (2011). http://doi.org/10.1088/2041-8205/728/2/L29
R.K. Ulrich, ApJ **172**, 165 (1972). http://doi.org/10.1086/151336
S. Vauclair, S. Théado, ApJ **753**, 49 (2012). http://doi.org/10.1088/0004-637X/753/1/49
F.C. Wachlin, M.M. Miller Bertolami, L.G. Althaus, A&A **533**, A139 (2011). http://doi.org/10.1051/0004-6361/201117029
F.C. Wachlin, S. Vauclair, L.G. Althaus, A&A **570**, A58 (2014). http://doi.org/10.1051/0004-6361/201424580
J.P. Zahn, in *Stellar Instability and Evolution*, *IAU Symposium*, vol. 59, ed. by P. Ledoux, A. Noels, A.W. Rodgers (1974), *IAU Symposium*, vol. 59, pp. 185–194
J.P. Zahn, A&A **265**, 115 (1992)

Chapter 3
External Mechanisms

External mechanisms are defined as those where the star obtains material from an external source. Planets and brown dwarfs (BD) are one plausible source of Li. This light element, sometimes used to distinguish brown dwarfs from low-mass stars (Rebolo et al. 1992), is burned in objects with masses higher than $M \sim 0.065\,M_{\odot}$ (Chabrier et al. 1996). The ingestion of this kind of objects could increase the superficial Li abundance of the star (Alexander 1967). Other possible sources of Li are binary companions. AGB stars can produce large amounts of Li in their interiors that due to convection increases the surface Li abundance of the star. Thus, if the giant has an AGB binary companion that transfers Li-enriched material, it could increase its Li abundance (Sackmann and Boothroyd 1999). Even if this process is an interesting alternative, only the higher-mass AGB are affected by the Li production process, and these are also found in a very specific metallicity domain.

3.1 Binary Companions

Possibly the only well-understood stellar source of high amounts of Li are intermediate-mass stars (4 to $7\,M_{\odot}$) in their AGB phase (Sackmann and Boothroyd 1992; Ventura and D'Antona 2010). As such, mass transfer from a binary companion of these characteristics could explain the high abundances found in RGB stars.

As the exact Li yield from AGB stars depends on processes still under discussion (e.g., mass loss), it is difficult to identify the maximum amount of Li that could be transferred, but given that AGB stars with A(Li) $>$ 4.0 have been found, it is possible that mass transfer from a Li-rich donor could produce RGB stars with A(Li) $=$ 3.3 or even higher. The mass transfer does not necessarily need to happen when the recipient star is on its RGB phase. If the star receives Li during the MS, even after dilution in the FDU, the Li abundance of the stars should remain high.

C. Aguilera-Gómez, *Explaining Lithium Enriched Red Giant Branch Stars*,
Springer Theses, https://doi.org/10.1007/978-3-030-02583-0_3

In this section we investigate if Li-rich RGB stars could be produced by mass transfer from an evolved massive companion, considering that the donor could have evolved into a white dwarf. To do this, we use data from the literature to compile a catalog of multi-epoch radial velocity measurements of Li-rich giants and look for possible binary companions based on the radial velocity monitoring.

It is interesting to notice that a detailed analysis of radial velocity variability of giants is usually done by authors when working with giants that are rotating unusually fast, some of them being also Li-rich giants. Although these particular determinations of radial velocity variability are relevant to identify possible binary companions that may produce the rapid rotation, this is not the type of binary companions that we want to focus on. These are closer binaries in shorter orbits. In contrast, we will attempt to identify possible companions with initial periods longer than $\sim$ 100 days, since those are the systems where the most massive star would reach advanced stages of evolution without transferring mass to the companion in previous stages (Thomas 1977). We need that the massive star reaches its AGB stage so it can produce Li, thus falling into the category of case C of mass transfer (when the mass transfer starts after helium exhaustion in the core).

de Medeiros et al. (1996) studied the radial velocities of some Li-rich giants and reported that there is no sign of binarity based on two measurements with a 1-year interval. Also, when studying both rotation and lithium abundance in single stars, De Medeiros et al. (2000) report several Li-rich giants with their respective rotation periods, none of which show signs of binarity. Other works also look for radial velocity variability of their target stars when looking for Li-rich giants, reaching similar conclusions, finding no signs of binarity in their target giants.

Here, we join catalogs of radial velocity by de Medeiros and Mayor (1999), De Medeiros et al. (2002), and Massarotti et al. (2008) and cross-reference these giants with the list of measured Li abundances from Rebull et al. (2015). Even if this last catalog includes Li-normal giants, it also contains most of the known Li-rich giants in the literature. Thus, we base our search on these stars. We find a total of 56 giants with different levels of Li abundance, 26 of which are Li-rich, with abundances A(Li) $\geq$ 1.5, that are part of the catalogs of radial velocities with at least 2 measurements in time baselines of more than 250 days.

Figure 3.1 shows the Li abundance and the probability that the radial velocity of the star is constant. This quantity is used as a powerful indicator of variability. It is the probability that the observed velocity variations of a star are due to observational errors. Low probabilities mean that there is a different source for the variability, which could indicate a binary companion. In other words, when stars show a probability close to zero, they are often binaries. Based on Gim et al. (1998), we show the vertical line at $P = 0.01$ that is used as a limit to probable binaries, although some works indicate that binaries have even smaller values (e.g., $P = 10^{-6}$, Carney et al. 2003).

Figure 3.1 shows that most of the stars, regardless of their Li content, do not seem to be binaries, based on observations in time spans close to a year or longer. There are no correlations between the Li abundance of the star and the binary status of the giant. If the phenomena of Li enrichment was somehow related to the presence of binaries, we could have expected to see a larger number of giants with probabilities

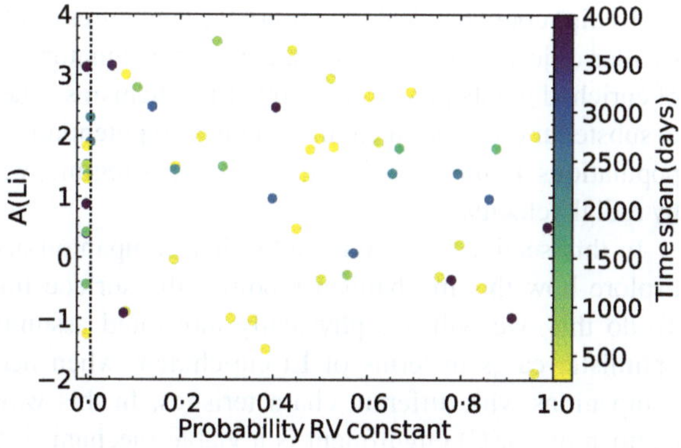

Fig. 3.1 Li abundance versus the probability that the radial velocity of the star is constant. They are color-coded according to the time baseline for observations that vary between 250 days and several years. Stars with probability values smaller than 0.01 (marked by the vertical dashed line) are often binaries

of constant radial velocity close to zero with high Li abundances, which is not the case. A strong limitation to this is that most giants here only have two measurements of radial velocity. To be more confident about their binary status, more data would be needed. However, for the purpose of Li enrichment by binary interaction, this confirms individual results by other works, indicating that a lot of enriched giants are isolated and have not been part of a mass transfer episode.

3.2 Substellar Companion Engulfment[1]

The interaction between stars and the planets they host is a complicated subject. It is known that when stars evolve to the RGB, their radii expand, leading to the possible engulfment of their orbiting planets, even showing signatures of a recent ingestion event (e.g., BD+48 740 Adamów et al. 2012). However, which planets will end up being accreted by the star depend on several physical variables included in models of tidal evolution, as well as the treatment of planetary orbits and interactions between star and planets (Villaver and Livio 2009; Kunitomo et al. 2011; Villaver et al. 2014).

Different mechanisms of Li enrichment could be producing giants in different regions of parameter space. Whether we want to study a specific enrichment process or we want to study the reason behind the enrichment for the entire population, we need to consider first which giants are appropriate for our studies. For instance, extra-mixing is expected to develop after the RGB bump, and thus we could expect Li-rich giants produced by this mechanisms with higher concentration in that phase of stellar evolution and afterward. As substellar companion ingestion is not limited to a particular evolutionary phase and it could develop before the luminosity function bump, a good approach to finding Li-enriched giants produced by engulfment is to look for signals in giants with a specific evolutionary phase, located after the end of the FDU but before the RGB bump. Additionally, populations that are known to have little to no lithium during the main sequence and subgiant branch phases (e.g.,

[1]Parts of this section were published in Aguilera-Gómez et al. (2018)

the Li dip) could be ideal to identify the signatures of planet engulfment, where even smaller objects could produce a detectable signature and there is no contamination of enriched giants produced by other mechanisms. The Li enrichment of giants due to substellar companion engulfment is also potentially interesting to study planetary populations around stars in a different mass regime, which is harder to characterize by radial velocity.

In this section, we model substellar companion accretion by the host star and explore how this mechanism modifies the surface lithium abundance of the star. To do this, we will use physically motivated assumptions focusing on the most optimistic cases in terms of Li enrichment when needed and consider stars and companions with different characteristics. In this way, we can study if substellar companion (SSC) engulfment is a viable mechanism to produce Li enrichment of giants, predict which giants are produced by this mechanisms, and thus see if SSC engulfment alone can explain the entire observed population of unusual objects. If this is not the case, our goal is to understand which subset of the observations can still be explained by this process.

The engulfment of SSCs has been modeled before in both the first-ascent red giants and asymptotic giant branch in Siess and Livio (1999a) and Siess and Livio (1999b), with a limited range of stellar masses, where they also explore other possible signatures of the accretion event, such as fast rotation and infrared excess. The stellar masses modeled in the RGB of around $1.0\,M_\odot$ are a limitation of the models, which we overcome with current tools, by using different initial conditions for stellar mass and metallicity and planetary composition, masses, and periods. This allows to study SSC engulfment in all domains and predict where in parameter space the signals of accretion are more easily and unequivocally detected. Combining the predictions from our models with knowledge of where different mechanisms act efficiently we can predict the Li abundance of giants where extra-mixing cannot operate. This, eventually, will lead to a better understanding of the Li abundance pattern of giants.

3.2.1 Models

In the SSC engulfment process, both the star and the substellar companion have to be modeled. Because of the difficulties associated with a self-consistent model of engulfment that includes the feedback of the process over the stellar structure and evolution, we will use a post-processing approach. From a technical perspective, this means that we use a grid of stellar evolutionary models as a starting point, including the substellar companion engulfment in a separate code, where the starting models of the star are used. This allows a faster exploration of the parameter space, which in this case is large and includes stellar mass, stellar metallicity, SSC mass, SSC composition, and the orbital period of accreted SSC. The approach used is useful; however, the stellar feedback could have some influence in the final Li abundance after engulfment in some specific cases.

Nevertheless, our approach is a reasonable starting point to better understand this problem and point out interesting regimes to study this problem or find candidates of planetary accretion, and it is a good starting point for more complicated models afterward. Additionally, this is a good approximation in general, given that the dredge-up of stars of different masses is similar, so the Li added by the SSC is a relatively small quantity that would mostly represent a perturbation to the FDU.

Star

The grid of canonical stellar evolutionary models needed (i.e., without rotation or overshooting) is obtained using the Yale Rotating Evolutionary Code (YREC, Pinsonneault et al. 1989, 1990, 1992; Demarque et al. 2008). These models use the mixing length theory for convection (Cox 1968), with 2006 OPAL equation of state (Rogers and Nayfonov 2002), nuclear reaction rates from Adelberger et al. (2011), and weak screening following Salpeter (1954). Given the importance of the Li abundance in our models, we update the cross section for the reaction ^{7}Li(p, α)α, which is reported by Lamia et al. (2012). Also, we use high-temperature opacities from the Opacity Project (Mendoza et al. 2007), complemented with Ferguson et al. (2005) low-temperature opacities. The atmosphere and boundary conditions are those from Kurucz (1979). van Saders and Pinsonneault (2012) presents a summary of the input physics used in our models.

Our grid of models, all evolved up to the tip of the red giant branch, covers a range of masses from $M = 1.0\,\mathrm{M}_\odot$ to $M = 2.0\,\mathrm{M}_\odot$, with increments of $0.25\,\mathrm{M}_\odot$. The composition adopted for these models considers a solar metal abundance $Z/X = 0.02292$ from Grevesse and Sauval (1998). The hydrogen mass fraction X and mixing length coefficient α are fixed to reproduce the solar radius and luminosity for a $1\,\mathrm{M}_\odot$ model at an age of 4.57 Gyrs considering gravitational settling. The calibrated values for these variables that we use in the models are $X = 0.70751$, $\alpha = 1.94661$, and $Z = 0.01879$.

For an arbitrary metallicity [Fe/H], the chemical mixture used in the model is obtained with a linear relation between helium mass fraction and metal content

$$Y = Y_p + \frac{\Delta Y}{\Delta Z} Z, \tag{3.1}$$

with $Y_p = 0.2484$ the primordial helium mass fraction (Cyburt et al. 2003) and $\frac{\Delta Y}{\Delta Z} = 1.3465$ the slope of the relation obtained by using both the solar-calibrated composition and the primordial mixture in Eq. (3.1).

The metallicities explored throughout this work are in the range [Fe/H]$= -2.0$ and [Fe/H]$= 0.0$ or chemical mixtures $Z = 0.00020$ and $Y = 0.2487$ to $Z = 0.01879$ and $Y = 0.2737$. Given that most of the observed Li-rich giants have metallicities closer to solar, the increment between metallicities is smaller at larger metallicities, with a much finer grid.

The initial Li abundance is an important initial condition in finding the maximum lithium abundance after engulfment of a SSC. The lithium depletion predicted by canonical stellar evolutionary models in pre-main sequence low-mass stars is smaller than the observed depletion when stars reach the turnoff. Therefore, we use a lower initial value of Li, adjusted to the maximum Li abundance observed among clusters (see Dobrovolskas et al. 2014, and references therein).

Given the Li abundances in the turnoff of clusters at solar metallicity, we consider three different mass regimes for the giants, each of these with a given initial Li abundance: stars with masses $M > 1.8\,M_{\odot}$ could have a higher Li abundance in the turnoff, similar to the meteoritic value of $A(Li)_{*,ini} = 3.3$. Low-mass stars with $M < 1.3\,M_{\odot}$, in contrast, correspond to an older population that would have a Li abundance at the turnoff from $A(Li)_{*,ini} = 2.3$ in the halo to $A(Li)_{*,ini} = 2.6$ at solar metallicity. There is also an intermediate range of masses, populated by stars in the Li dip that leave the MS with very low Li abundances, where the abundances are mainly upper limits $A(Li)_{*,ini} < 1.0$.

To identify how the Li abundance after SSC engulfment depends on parameters such as mass and metallicity without the added complication of different initial lithium abundance, we also produce models with an initial Li abundance of $A(Li)_{*,ini} = 2.6$, regardless of the mass of the star.

Substellar Companion

Current stellar models are capable of modeling low-mass objects, similar to the SSCs that we will consider in this work. The use of such detailed calculations is beyond the scope of this work, and we consider it to be unnecessary. To explore the upper envelope of the Li distribution signature after engulfment, the only parameters considered are mass, radius, and composition (including the amount of Li) for the planets and BDs.

The mass, radius, and composition of planets are obtained from the grid of Fortney et al. (2007), for both giant planets and rock-ice-iron planets. For brown dwarfs, instead, we use the mass and radius obtained with evolutionary models of Baraffe et al. (2003) and assume a composition similar to that of Jupiter and Saturn (hydrogen and helium mainly). The masses of accreted objects go from $1\,M_{\oplus}$ up to $0.065\,M_{\odot}$, where BDs of higher masses are not considered, as they would burn the Li during their lifetimes (Chabrier et al. 1996), being unable to increase the Li content of the star.

Changing the specific model used to describe the SSC does not affect our final result. No differences were observed when modeling rock-iron-ice planets using the grid of Zeng and Sasselov (2013).

SSC Li Content

The Li abundance of the SSC is an important parameter for the maximum Li after SSC engulfment. Thus, we need the total lithium mass available in the SSC that will be accreted by the star. To derive this, we fix the ratio between Li mass fraction and metals to the solar system meteoritic value. This will assume that all SSCs have the same X_{Li}/Z but may have a different metal content Z_{SSC} and, consequently, a different mass fraction of Li. Then, we will just need the mass of the SSC to obtain the accreted Li content.

The meteoritic Li mass fraction was calculated by using an abundance of A(Li) = 3.3 (Lodders and Fegley 1998). Exact planet composition, in particular regarding the light element Li, is uncertain and still debated, even for planets inside the solar system. For our models, we consider three mass domains for the SSC companion, where we choose the metal content accordingly. Brown dwarfs are expected to have relatively high Li abundance when compared to stars as this light element can be used to distinguish between substellar objects and very low-mass stars (the "lithium test" Rebolo et al. 1992). Although we expect BD to have a significant amount of lithium, in every other sense, they have a similar composition to stars. Thus, to characterize them, we will use the primordial solar mixture, with a value for the overall planetary metallicity of $Z = Z_{\odot}$. We also model the engulfment of BD enriched in metals, given the possibility that they may be formed by mass accretion. We use this composition for SSCs with masses $15M_J < M_{SSC} \leq 60M_J$.

Our knowledge on the composition of planets does not allow to precisely identify their Li abundances. Based on what is found in the solar system, the rocky planets tend to be overabundant in heavier, higher-condensation temperature elements, while the giant gaseous planets share a similar composition to their parent star (e.g., Palme 2000). Given this apparent trend, we can anticipate the Li enhancement that could be produced by the engulfment of a Jupiter-like planet. In spite of its large mass, a gaseous planet with its small amount of Li could be insufficient to increase the stellar Li to the high amount found on the surface of Li-rich giants. On the other hand, the rocky planets that have a larger Li reservoir than gaseous giant planets are less massive, and thus the signature they produce in the star could be too small to detect.

Considering these possible differences, we will use the same Li abundance for all SSCs modeled but separate planets into different mass domains, rocky and giant planets, which will have different metal content. Giant planets with masses from $0.01\,M_J$ to $15\,M_J$ will have $Z_{SSC} = 2.5\,Z_{\odot}$ based on what is found in Jupiter (Nettelmann et al. 2012), where these planets are slightly more metal-rich than their parent star because of their formation by core accretion. Rocky, lower-mass planets, including Earth-like planets and all other companions with $M_{SSC} < 0.01\,M_J$, are considered to be enriched in metals, with $Z = 1$.

Assumptions

In this section, we specify and describe the assumptions used to model the ingestion of the SSC.

We do not consider the tidal evolution of the SSC before it is engulfed by the star, a process that is still debated and discussed in the current literature and that is not relevant for the purposes of Li enrichment. Instead, we find a point of planet dissipation in the interior of the star, considering the characteristics of both the planet and the star (see Sect. 3.2.2). This assumption simplifies the calculations, given that we ignore the orbit of the SSC prior to the point of dissolution, but it still allows us to consider planets with different main sequence orbital periods (i.e., different distances to their parent star).

Our second assumption is that the SSC is completely dissolved in that specific point inside the star, without any accretion rate or mass loss of the SSC before it reaches the point of dissolution. To know which portion of the planet or BD is dissolved in the convective envelope of the star, knowledge of the mass of the companion and its detailed structure is needed (Sandquist et al. 1998). Part of the accreted companion may be evaporated during its orbital evolution, before arriving at the point of complete dissolution, or even before there is any contact between the planet and star. The Li mass accreted in the case that the SSC is partially and gradually dissolved inside the star is the same than in the case where the companion dissolves in one point inside the convective envelope. If the planet or brown dwarf is stripped of mass before contacting the star, that mass would not contribute to increasing the Li enrichment. As such, our assumption would always lead to the best case scenario in terms of Li abundance of the star.

Regarding the convective region of the star, the timestep used in our models, corresponding to the nuclear timescale, is longer than the convective turnover timescale. Thus, we can safely assume that the convective regions are chemically homogeneous, mixing instantaneously the chemical elements.

At last, as was mentioned previously, we use a post-processing approach, which assumes that the engulfment of the SSC does not affect the internal structure and evolution of the star.

3.2.2 Point of SSC Dissipation

To model the Li abundance of the star after the engulfment of the SSC, we require knowledge of the point where the planet dissolves inside the star, i.e., where it will deposit its Li. In the case the SSC is dissolved in the radiative interior of the star, no observable signature would be produced in the surface of the star, considering that no mechanism of extra-mixing is connecting the radiative region with the convective envelope.

Based on Siess and Livio (1999a) we find the point where the SSC will dissolve inside the star, calculating the radial distance from the center of the star where the ambient temperature is comparable to the virial temperature of the SSC:

$$T_{\rm vir} \approx \frac{G \mu_{\rm SSC} m_H M_{\rm SSC}}{k R_{\rm SSC}}, \tag{3.2}$$

where the *SSC* index designates the different characteristics of the SSC. M is the mass, μ the mean molecular weight, and R the radius. The physical quantities gravitational constant, Boltzmann constant, and hydrogen mass are G, k, and m_H, respectively.

This point inside the star, where we assume the SSC is completely dissolved, is where the kinetic energy of the star is similar to the gravitational binding energy of the external object. Once again, this assumption conserves the Li deposited in the star by the SSC, as the dissipation in inner regions of the star, where the temperature is higher, could destroy the deposited Li.

The virial temperature characterizes the SSC, as it depends on its specific properties. However, different SSCs can have the same virial temperature and, as such, will evaporate at the same point inside a given star. Notice that we find the region inside the star where the SSC is dissipated by comparing the virial temperature of the SSC with the temperature profile of the star, which means that different stars, with different temperature profiles, will dissolve the SSC in a different location.

Some conditions can change the locus of SSC dissipation. In particular, during the orbital evolution of the SSC, some of the kinetic orbital energy can be converted into thermal energy, changing the temperature profile of the star and moving the point of dissipation closer to the stellar surface. Although we do not consider the orbital evolution of the SSC, one of the limitations of the models, the difference that could be produced in the dissipation point is not relevant in the Li enrichment, given that it does not modify the final abundance. With an order of magnitude calculation, we find that the effect of kinetic orbital energy is practically negligible for the SSC modeled in this work, and it is only relevant for higher-mass objects.

Tidal forces that act inside the star can also be important, disrupting the SCC. For stars, the tidal effects are more important as the SSC approaches the stellar core (Soker et al. 1987). Comparing both criteria of dissolution, we find that the point of disruption based on the virial temperature is always closer to the stellar surface. Thus, when the tidal forces are important, the SSC has already been dissolved inside the star. Moreover, if the SSC were dissolved at the inner radius that tidal forces predict, the external material would be deposited in the radiative region of the star, unable to produce an observable signature in the surface of the star in the absence of extra-mixing processes connecting the convective with the radiative zone.

The actual destiny of the planet inside the envelope of the star, with no assumptions about instantaneous dissolution, is given by the common envelope evolution of the system, a process not fully understood quantitatively. Hydrodynamical models of planet engulfment show the dissipation of the planet through heating while it is inside the convective envelope. Staff et al. (2016) mention how their simulated planet is not disrupted by shear instabilities as it spirals in toward the star, and instead, it dissolved in the convective envelope by heating. Although higher-mass objects could survive the common envelope phase (e.g., Nelemans and Tauris 1998), lower-mass companions could dissolve inside the star.

In this work, we model the effects of engulfment of SSCs in the surface composition of giants, but the accretion of companions could also have important effects on the stellar rotation due to the transfer of angular momentum (Privitera et al. 2016).

Table 3.1 Examples of planets and brown dwarfs with their calculated virial temperatures. Originally in Aguilera-Gómez et al. (2016b)

Type of SSC	Mass (M_J)	Radius ($R_\oplus$)	Composition	Virial temperature (K)
Planet	1.42×10^{-4}	0.48	Rock/Iron	7.609×10^3
Planet	3.15×10^{-3} [a]	1.00	67% Rock/33% Iron	7.089×10^5
Planet	5.35×10^{-2}	7.26	Hydrogen/Helium	3.543×10^4
Brown dwarf	20	10.45	Hydrogen/Helium	9.644×10^6
Brown dwarf	60	8.69	Hydrogen/Helium	3.479×10^7
Planet	1.46	12.96	Hydrogen/Helium	6.067×10^5
Planet	1.00×10^{-3}	0.77	Rock	4.423×10^5
Planet	3.15×10^{-3}	0.77	Iron	5.486×10^5

[a]Equivalent to $1.00\,M_\oplus$

Using models for the composition of planets and brown dwarfs as indicated previously in this section, we show some example SSCs in Table 3.1 with their properties and virial temperatures. We also note that a Jupiter-like planet would have a very similar virial temperature to that of an Earth-like planet.

Figure 3.2 shows the point where four different SSCs (with their specific virial temperatures) dissolve inside $1.0\,M_\odot$ stars of different metallicities. These four modeled objects cover a wide range of virial temperature values which can be derived from companions with very different properties. Each gray line is the point of dissipation of a specific SSC. Additionally, we add gray bands around that line showing a $\sim$ 5% value around each virial temperature in logarithm. The black dashed line shows the separation between convective and radiative regions of the star.

Comparing the four different panels in Fig. 3.2, we see that, at a given moment in stellar evolution, the point of dissipation is similar for stars of different metallicities. This figure also shows interesting structural properties of stars and how they change with stellar properties, which impacts directly on the Li surface abundance of canonically evolving giants. The larger opacities of metal-rich stars difficult the radiative transport inside them, thus we find that the growth of the surface convection zone is much larger in these giants. Figure 3.3, with the locus of SSC dissipation inside $1.5\,M_\odot$ stars, indicates that this trend is independent of stellar mass.

It is relevant to notice that it is usually reported that Li is burned at a temperature of 2.5×10^6 K. Although this is true for MS stars, the conditions required for burning this element are not the same during the RGB due to the lower densities found inside the star. Thus, even if the convective envelope deepens more, reaching hotter temperatures in higher-metallicity stars, we do not find a higher rate of Li burning inside the star given its properties.

Understanding whether SSCs dissolve in the radiative or convective region is key in finding which objects can produce an observable signature in the surface of the star. To clarify this, Fig. 3.4 depicts the ambient temperature found in the bottom of the convective envelope, separating the convective region (shaded) from

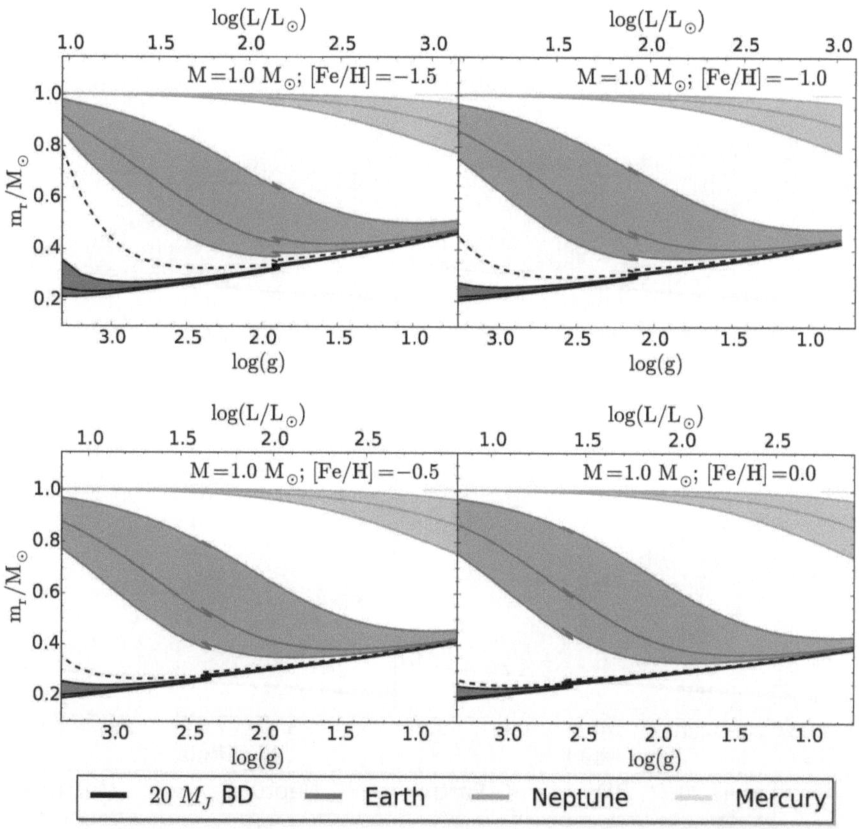

Fig. 3.2 Locus in mass inside the star where four different SSCs would dissipate during the RGB phase of a $1.0\,M_\odot$ star. The evolution of the star is both shown in luminosity and $\log g$. At the top of each panel, we find the specific metallicity of the modeled star. Gray bands (with 5% errors at each side) indicate the dissolution point for SSCs as indicated in the legend at the bottom of the figure. The black dashed line shows the base of the convective envelope. For different metallicities, the SSC dissolution point is very similar at each moment in the evolution. Also, SSCs with larger virial temperatures dissolve in the radiative region of the star. Credit: Aguilera-Gómez et al. (2016b)

the radiative region of stars of different masses and metallicities. As an example, we also show the virial temperature of a $15\,M_J$ brown dwarf with a dashed line. From this figure, we see that planets that have a higher virial temperature will dissolve in the radiative interior of the star. These correspond to the more massive SSCs or those with smaller radii. These particular objects cannot produce a visible signature in the star, and only the extra-mixing mechanisms not modeled in this work would allow the Li deposited in the interior to be transported to the stellar surface.

Thus, we focus on SSCs that dissolve in the convective envelope and that can produce a signature. The example brown dwarf in Fig. 3.4 dissolves in the convective region for higher metallicities and in the radiative interior for lower

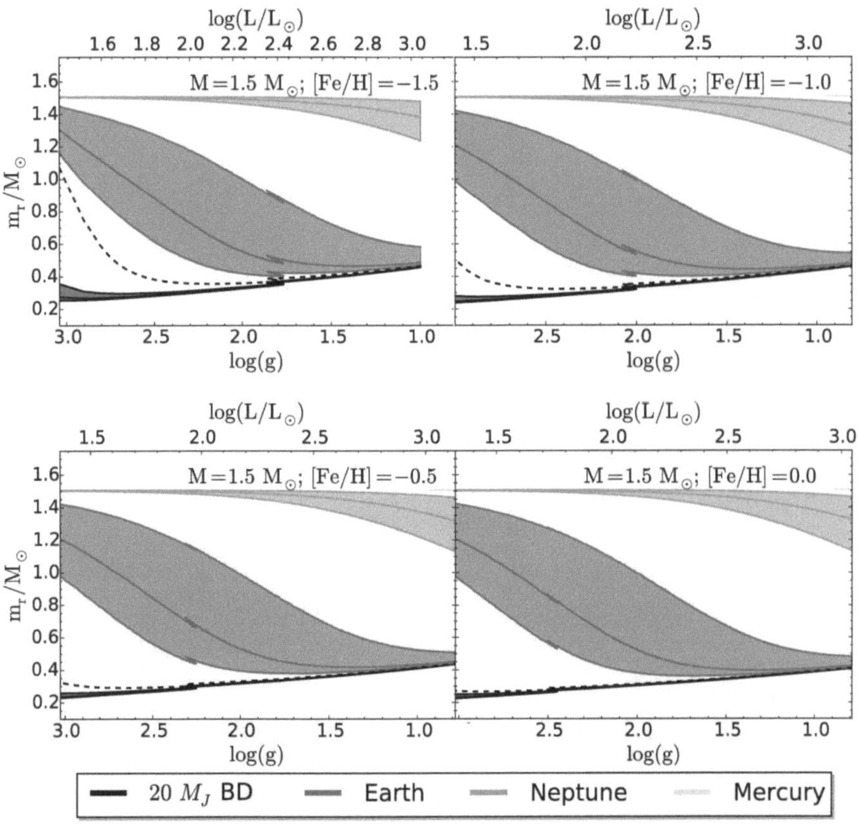

Fig. 3.3 Same as Fig. 3.2 with a parent star of 1.5 M$_\odot$ stars. Credit: Aguilera-Gómez et al. (2016b)

metallicities. This does not only show that there is a metallicity dependence on the SSC that could produce an observable signature in the star but also that stars with different properties can engulf and dissolve different objects in their lives.

Additionally, the brown dwarf shows a limiting case for a SSC that always dissipates close to the convective base. Objects much more massive will dissolve in the radiative interior. As such, objects of masses 15 to 20 M$_J$ work as an upper bound of companions that would dissolve in the convective envelope, depending on the exact metallicity of the star. This limiting mass leads to conclude that, although we could have expected that SSCs with larger mass had a larger reservoir of Li to contribute to the star, they dissolve in the radiative interior. Instead of producing the largest possible Li abundance post-engulfment they would disappear in the radiative interior without leaving trace of the process in the surface Li abundance of the star.

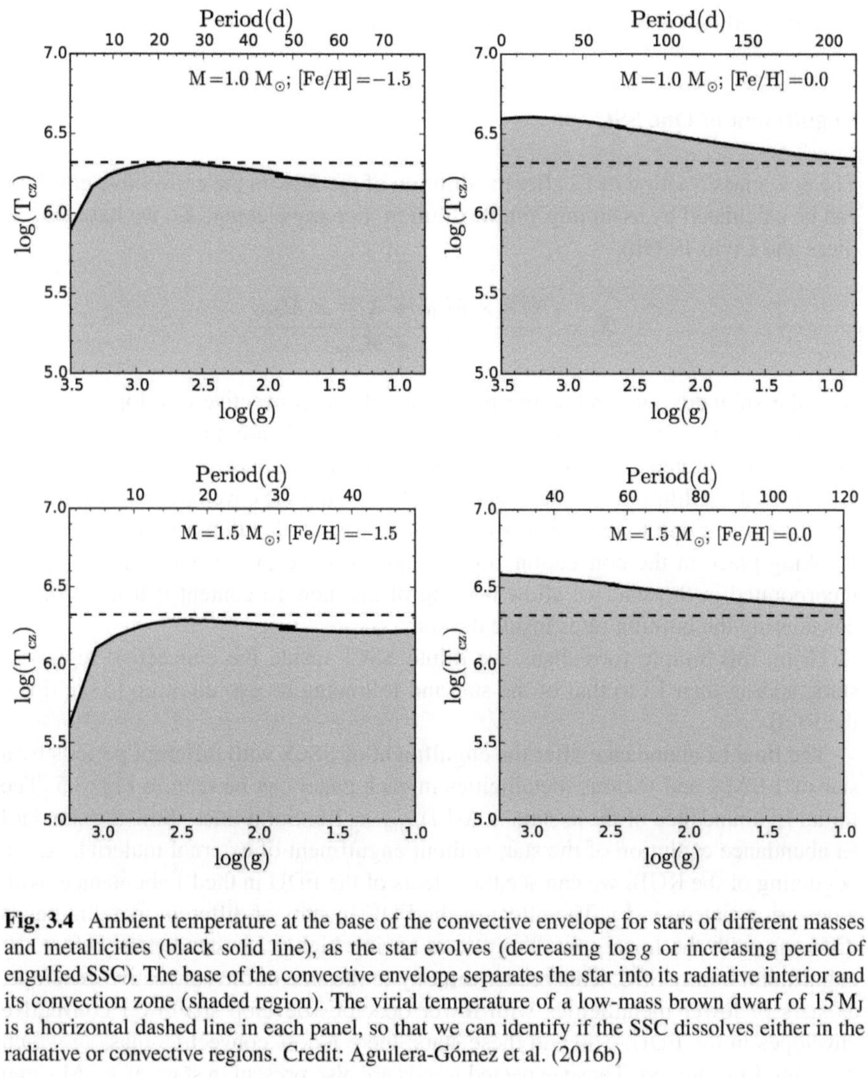

Fig. 3.4 Ambient temperature at the base of the convective envelope for stars of different masses and metallicities (black solid line), as the star evolves (decreasing $\log g$ or increasing period of engulfed SSC). The base of the convective envelope separates the star into its radiative interior and its convection zone (shaded region). The virial temperature of a low-mass brown dwarf of $15\,M_J$ is a horizontal dashed line in each panel, so that we can identify if the SSC dissolves either in the radiative or convective regions. Credit: Aguilera-Gómez et al. (2016b)

3.2.3 Surface Li Abundance

In order to see how the planet engulfment process affects stars of different properties, we first model the engulfment of one SSC. In particular, we choose the higher-mass object that would dissolve in the convective envelope of stars, a low-mass brown dwarf (or high-mass planet) of $\sim 15\,M_J$ and $Z_{SSC} = 2.5\,Z_\odot$. After understanding the effects of this particular object, we show results for SSCs with different properties. Throughout this section, and when it is not stated otherwise, the initial lithium abundance used for stars is $A(Li)_{*,ini} = 2.6$.

Engulfment of One SSC

The new mass fraction of Li after the dilution of the SSC in the convective envelope can be calculated by assuming simple dilution. For any element X_i, we have (Eq. 2, Siess and Livio 1999b):

$$X_i = \frac{X_i^{env} \times M_{env} + X_i^{acc} \times M_{acc}}{M_{acc} + M_{env}}, \tag{3.3}$$

with the subindex *env* indicating properties of the convective envelope and *acc*, properties of the accreted SSC. M are masses, while X are mass fractions. In our models, the accreted mass M_{acc} corresponds to the entire mass of the SSC. Here, we assume that the dilution is instantaneous, not changing the envelope mass during the accretion event due to canonical RGB evolution, and also that no nuclear burning is taking place in the convection zone. However, once the Li mass of the SSC is incorporated in the star, we allow burning of this new Li content if it is adequate, considering the burning rates inside the star.

Using this simple formalism, we dilute SSCs inside the convective region of stars, adding their Li to that of the star and following its evolution up to the tip of the RGB.

The final Li abundance after the engulfment of SSCs with different periods by a star of $1.0\,M_\odot$ and various metallicities in each panel can be seen in Fig. 3.5. The initial Li abundance of these stars is $A(Li)_{*,ini} = 2.6$. Solid lines show the standard Li abundance evolution of the star, without engulfment of external material. At the beginning of the RGB, we can see the effects of the FDU in the Li abundance, with a posterior plateau in Li. The effect of the FDU in stars of different metallicities is different, with the lower-metallicity stars having higher Li abundances before the engulfment of any SSC. This relates directly to the size of the convective envelope, as stars of lower metallicities with lower opacities develop shallower convective envelopes in the FDU. Thus, in these stars, there is less convective mass in which the same Li is diluted. These expected trends are also present in stars of $1.5\,M_\odot$ and $2.0\,M_\odot$ (Figs. 3.6 and 3.7). Although the stars with lower metallicities tend to have higher Li abundances, the effect is more important in lower mass objects.

If we now consider the effect of the accreted SSC on the A(Li), we see that the engulfment produces a rapid increase in Li as soon as the Li content of the planet is diluted in the convective envelope. If the engulfment is produced previous to the FDU, we can also see the effect of this process on the total Li abundance of the giant including that contributed by the SSC. Standard dependencies in stellar parameters such as the depth of the FDU as a function of mass and metallicity are preserved in a post-processing approach.

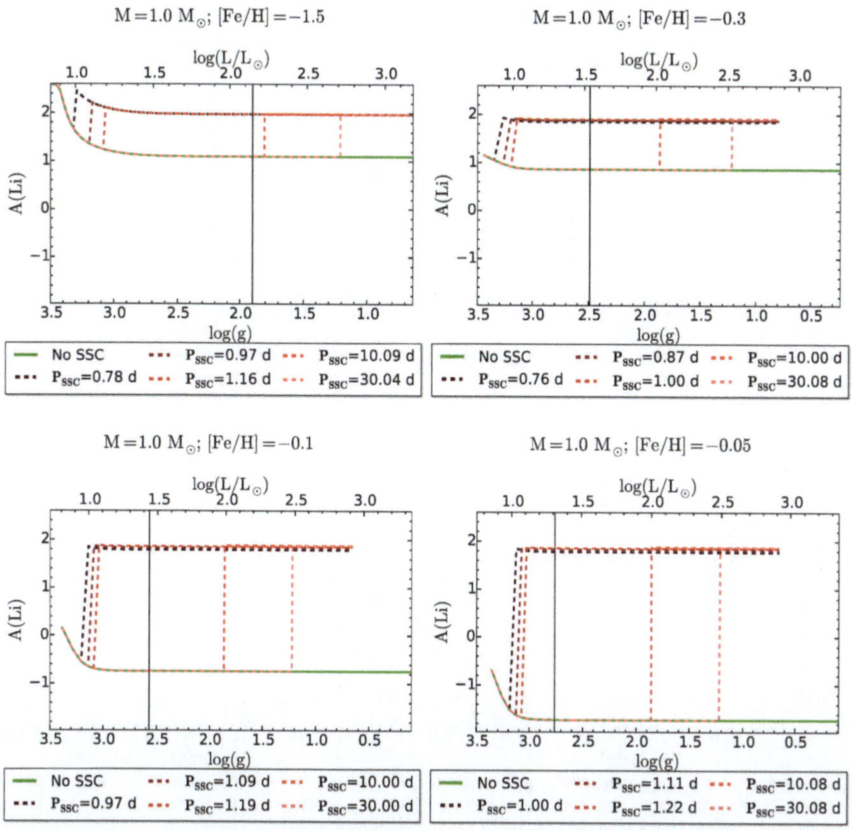

Fig. 3.5 Surface Li abundance in a 1.0 $M_\odot$ star with initial lithium abundance of $A(Li)_{*,ini} = 2.6$. Each panel shows a different stellar metallicity. Solid lines are the Li abundances A(Li) for the star evolving canonically, without engulfment. The dashed lines are A(Li) after engulfment of SSCs with different periods P_{SSC} in days. The evolution of stellar Li is shown with both $\log g$ and luminosity. The luminosity function bump is indicated as a solid vertical line in each panel. Credit: Aguilera-Gómez et al. (2016b)

With such a large SSC in mind, the Li abundance in every case increases significantly. The amount of enrichment depends on the properties of the star. Lower-metallicity stars display the higher abundance after engulfment because they have the largest Li abundance prior to accreting the SSC. However, the higher-metallicity stars are those presenting the largest difference between lithium abundance prior to and after the engulfment. An example of this is the 1.0 $M_\odot$ star with [Fe/H]$= -0.05$. With a canonical Li abundance of A(Li) ~ -1.7, the star can reach a value of A(Li) $\sim$ 1.9 by engulfing a SSC with period $P_{SSC} = 1.42$ days, increasing significantly its Li compared to the initial abundance.

Alongside dilution, another process to be considered when analyzing the Li abundance is nuclear burning. We allow the burning of planetary Li inside the star

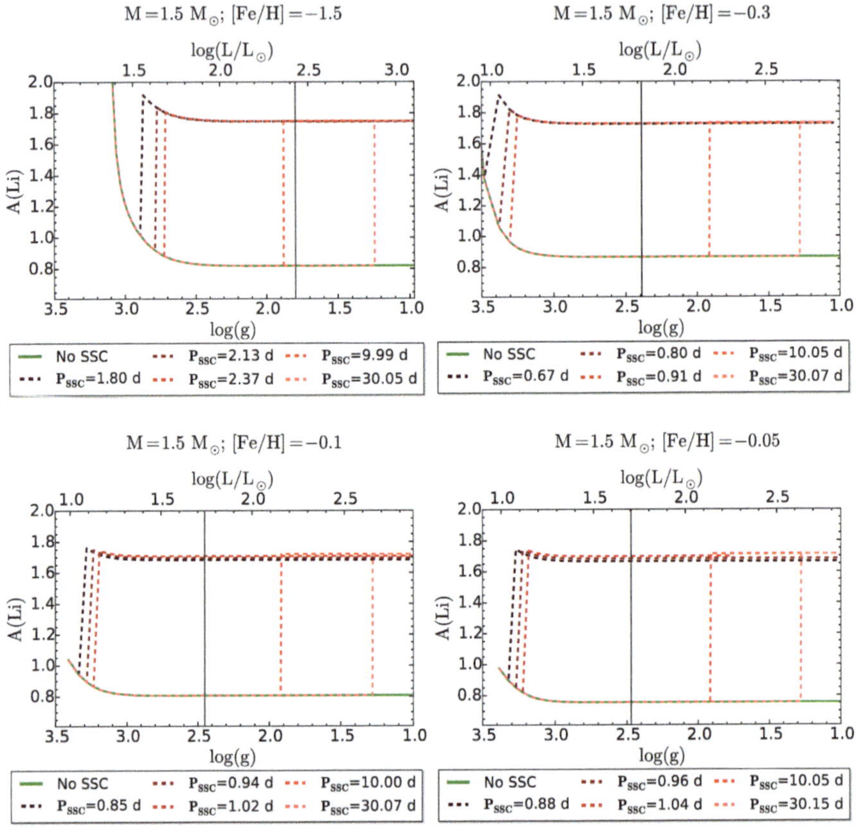

Fig. 3.6 Same as Fig. 3.5 for a 1.5 M$_\odot$. Credit: Aguilera-Gómez et al. (2016b)

according to the nuclear reaction rates calculated. If conditions are favorable for Li burning in convective conditions, there could be a difference between the Li abundance right after engulfment and time after. We notice again, however, that even if the temperature of the base of the convective envelope can be higher than $T \sim 2.5 \times 10^6$ K, this does not imply burning of Li, as the densities in the RGB phase are lower.

Burning of planetary Li does not affect the lower-metallicity stars in any point of their evolution, and as such, we see that planets with different periods engulfed by the star contribute the same Li to the star after long periods of time, or, in other words, at the tip of the RGB, Li abundance post-engulfment is the same, regardless of the period of the engulfed planet. The only thing that changes, in this case, is the $\log g$ where the initial sudden increase in Li is produced.

In contrast, for higher-metallicity stars, there is burning of Li in the convective envelope during FDU, and thus, part of the Li that previously belonged to the SSC is immediately burned after the engulfment. This implies that the total Li contribution

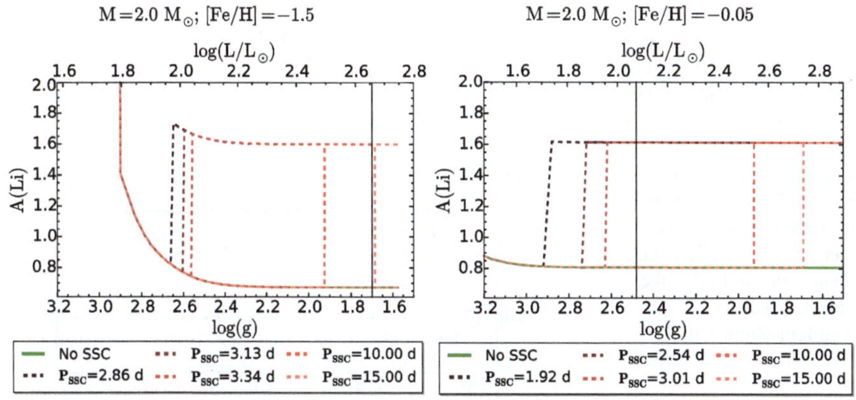

Fig. 3.7 Surface Li abundance of a 2 M$_\odot$ host after engulfment. Symbols are the same as in Fig. 3.5. In this mass range, the final Li abundance does not change with period of engulfed planet regardless of metallicity, so we show only two panels with different metal content. Credit: Aguilera-Gómez et al. (2016b)

by the SSC is smaller than what is expected by simple dilution of the SSC content on the convective envelope. This is the reason behind the small differences in Li abundance depending on the period of the engulfed SSC. Concentrating on the higher-metallicity panels of Fig. 3.5, we see that SSCs with smaller periods, engulfed previous to the end of the FDU, produce a slightly smaller increase in the Li abundance because the higher densities found in the bottom of the convective envelope allow the burning of SSC Li. Once the FDU ends and the convective envelope reaches its maximum depth, no further Li burning is expected and the Li after this point remains mostly constant, only decreasing by an extremely small amount.

In Figs. 3.5, 3.6, and 3.7, we see that the Li abundance does not change if SSCs are accreted after the FDU, because of the lower rates of Li burning. Thus, by considering the engulfment of SSCs after the end of the FDU, we can place an upper bound on the Li abundance that can be obtained with planet engulfment.

Stars of 1.5 M$_\odot$ of higher metallicity also show this effect related to the burning of the Li provided by the SSC. However, on stars of higher mass (Fig. 3.7), the Li abundance does not depend on the period of the engulfed planet or the time of accretion, and it only depends on the intrinsic properties of planet and star.

So far, we have analyzed one specific SSC of 15 M$_J$. The maximum Li abundance post-engulfment is produced in a star of 1.0 M$_\odot$ and [Fe/H]= −2.0 and corresponds to a value of A(Li) ∼2.0. This star presented an increase of 0.9 dex over its canonical Li abundance after accreting the SSC. Even though the maximum Li abundance is obtained in this star, the higher increase can be seen in lower-metallicity giants, reaching 4.0 dex of difference between the canonical star and that with engulfment. A large increase does not imply a higher final Li abundance. In particular, if the canonical abundance of the star is low, it would only be measured as an upper limit in A(Li), and in consequence, the considerable increase in Li after engulfment could not be measured.

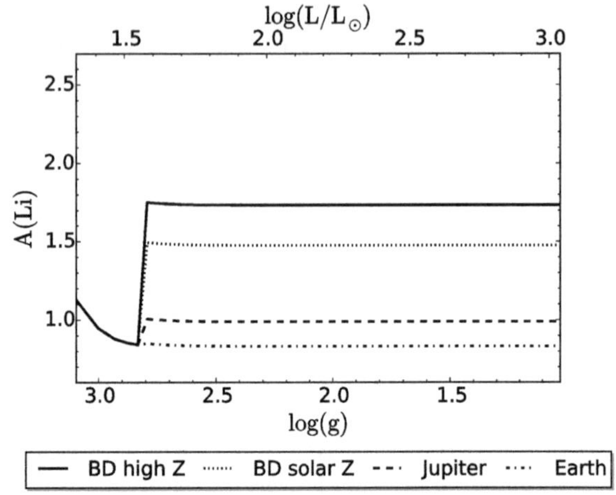

Fig. 3.8 Surface Li abundance in 1.5 $M_\odot$ stars with [Fe/H]=−1.0 that engulf four different SSCs at the same time. Engulfed SSCs are, as indicated in the legend at the bottom, a brown dwarf with high metal content, a brown dwarf with solar metallicity, a Jupiter-like planet, and an Earth-like object. Credit: Aguilera-Gómez et al. (2016b)

Changing the Engulfed SSC

We now model four realistic planets with different intrinsic characteristics (masses and compositions, including their Li abundance): the first of these SSCs is the same modeled in the previous section, a 15 M_J brown dwarf with Jupiter-like composition ($Z_{SSC} = 2.5\,Z_\odot$, formed by core accretion). The other modeled SSCs are a Jupiter-mass planet with $Z_{SSC} = 2.5\,Z_\odot$, an Earth-like planet with rocky composition ($Z = 1.0$), and a brown dwarf of 15 M_J with solar metallicity. We note that higher-mass SSCs are not considered because they would dissolve in the radiative interior not producing a visible signature in the surface of the star.

Figure 3.8 shows the difference in final Li abundance post-engulfment in a star of 1.5 $M_\odot$ with [Fe/H]= −1.0 when it engulfs four different SSCs in the same timestep. The highest enrichment is produced by the brown dwarf enriched in metals that we modeled in the previous section, followed by the brown dwarf with solar metallicity, with a Li enhancement only $\sim$ 0.3 dex lower than the BD enhanced in metals. The Jupiter-like planet increases the Li abundance of the star by only $\sim$ 0.2 dex. Finally, although Earth-like planets have the higher-mass fraction of Li of all modeled SSC due to their higher metal content, given their low masses they produce almost no enhancement in the Li abundance of the star.

Changing the Starting Li Abundance of the Host Star

In previous sections we have considered only giants with initial Li abundances of $A(Li)_{*,ini} = 2.6$. Now, we will consider host stars with two additional different starting Li to explore the known range of Li abundances of MS stars, $A(Li)_{*,ini} = 3.3$, the meteoritic value, and $A(Li)_{*,ini} \ll 1.0$, considering stars that have evolved from the Li dip and leave the MS with very low Li abundances. Some catalogs of

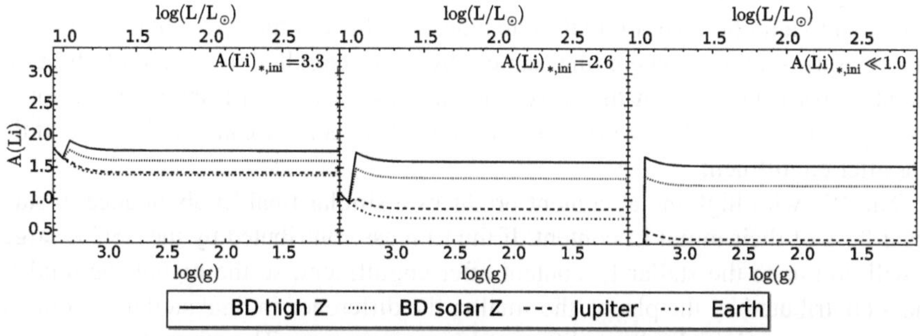

Fig. 3.9 Surface Li abundance for 1.5 $M_\odot$ stars with solar metallicity and different initial stellar Li abundances before and after engulfment of SSCs. From left to right panel, the initial Li is $A(Li)_{*,ini} = 3.3$, $A(Li)_{*,ini} = 2.6$, and $A(Li)_{*,ini} \ll 1.0$. In each panel, we show results for four engulfed SSCs: a BD with enhanced metals, a BD with solar metallicity, a Jupiter-like planet, and an Earth-like planet, with an orbital period in main sequence of $P_{SSC} = 0.78$ days. In the rightmost panel, the Li abundance post-engulfment of an Earth-like planet is $A(Li) = -0.7$. Credit: Aguilera-Gómez et al. (2016b)

Li abundance in dwarfs (Ramírez et al. 2012; Delgado Mena et al. 2015; Aguilera-Gómez et al. 2018) show the wide range of abundances found in the MS.

The effects of changing the initial Li abundance of stars is shown in Fig. 3.9. Here we present a 1.5 $M_\odot$ host star with solar metallicity and different initial Li content. From left to right, we show abundances of $A(Li)_{*,ini} = 3.3$, $A(Li)_{*,ini} = 2.6$, and $A(Li)_{*,ini} \ll 1.0$. We modeled the engulfment of the same four planets discussed previously (BD with enhanced metals, BD with solar metallicity, Jupiter-like, and Earth-like), with a period of $P_{SSC} = 0.78$ days.

Regardless of the initial Li abundance of the star, the BD with enhanced metals always produces the largest Li enrichment, followed by the solar metallicity BD, the Jupiter planet, and the Earth planet, this last planet with such a small enhancement that is barely detectable. We do not show in the rightmost panel the extremely low Li abundance after engulfment of an Earth-like planet, which corresponds to a $A(Li) = -0.7$. Although the difference between the canonically evolving giant and a star that accretes an Earth-like planet is important, the initial abundance is so low, that even this large signature would remain undetected.

In stars with a higher initial Li reservoir, $A(Li)_{*,ini} = 3.3$, the effect of planet engulfment is much lower because a much larger Li needs to be diluted in the same convective envelope to produce a visible increase. And with such a high abundance, there is less difference in the Li abundance post-engulfment for the four different SSCs.

Focusing only on how the BD with enhanced metals changes the Li content of the star in the different panels of Fig. 3.9, we see that even if the Li provided by the SSC is the same, a more significant enhancement is seen in canonical stars with less Li. Also, the large difference in Li between stars prior to the engulfment is decreased after the accretion of the SSC, because the same SSC dissolved in a region with more Li produces a smaller signal.

A larger difference between the canonical giant and that with engulfment does not equal to a higher overall final Li abundance. This is similar to what we modeled in previous sections, where the same SSC was engulfed by stars of different metallicities (Fig. 3.5). In that case, the star with a lower Li abundance prior to the engulfment had a larger difference in A(Li) between the canonical giant and the star after engulfment.

The BD with high metal content produces a similar final Li abundance in stars regardless of their initial Li content. If the Li mass contributed by the SSC is large, it will dominate the stellar Li content after engulfment, so the higher the total Li mass contributed by the planet, the smaller the difference in final stellar Li content, even if the stars initially have different Li compositions. Therefore the abundance post-engulfment when the SSC has a large Li content is similar in different stars, regardless of their $A(Li)_{*,ini}$.

The largest Li enhancement obtained exploring the entire parameter space can be found in a $1.0\,M_{\odot}$ star with [Fe/H]=−2.0, and $A(Li)_{*,ini} = 3.3$, after the accretion of a BD with enhanced metal content. This large abundance corresponds to A(Li) = 2.2, as we see in Fig. 3.10.

The same effects of Fig. 3.9 are replicated in Fig. 3.10.

We will now define which SSCs can produce an observable signature in different stars. For this purpose, we consider three different mass domains as outlined in section "Star", with different initial Li abundances. Low-mass stars have an abundance of $A(Li)_{*,ini} = 2.6$, but we also consider intermediate-mass stars evolving from the Li dip with $A(Li)_{*,ini} \ll 1.0$ and higher-mass stars with $A(Li)_{*,ini} = 3.3$. For this calculation, we take a star of different metallicities and find the maximum possible Li abundance of canonical giants, and SSCs which produce a visible signal are only those that enrich the giant above that maximum canonical Li. For the giants evolving from the Li dip, any planet could produce a higher signature than the canonical Li of the star. As such, we choose two thresholds

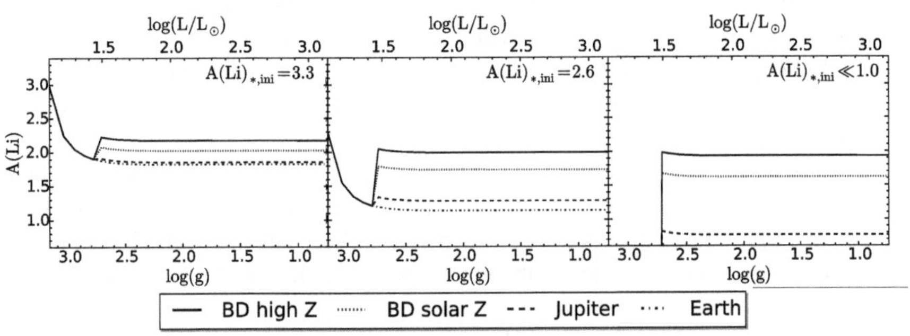

Fig. 3.10 Surface Li abundances of stars of $1.0\,M_{\odot}$ and [Fe/H]=−2.0 with different initial Li content (as indicated at the top of each panel). Symbols for the engulfed SSC are the same as in Fig. 3.9. The period of the engulfed SSC is $P_{SSC} = 2.00$ days. In the rightmost panel, the accretion of an Earth-like planet produces a stellar Li abundance of A(Li) ∼ −0.3. Credit: Aguilera-Gómez et al. (2016b)

Table 3.2 Mass of the SSC that can produce a visible signature after engulfment in stars of different masses. A visible signature is defined as the Li abundance threshold needed so that the signal is higher than the Li content of the canonical giants. Credit: Aguilera-Gómez et al. (2016b)

Mass range	A(Li) threshold	Planet mass (M_J)
Low-mass stars	1.25	6.7
Evolved Li-dip stars	0.5	2.6
Evolved Li-dip stars	1.0	8.6
High-mass stars	1.52	6.0

to define what is a detectable signature, a limit of A(Li) = 0.5 and A(Li) = 1.0. Table 3.2 presents the minimum planetary mass needed to produce a signal in the giants.

Our definition of visible signature implies that low mass SSCs can still produce an increase in the Li content of the star. However, the values in Table 3.2 indicate the minimum mass of companions needed to produce a signal higher than the Li of canonical giants, and as such, if the abundance is that high, it must be produced by a nonstandard mechanism such as planet engulfment.

One last note about the enrichment of an individual star engulfing SSC is that the increase in Li is linear, and if more than one SSC is accreted, they would produce an effect similar to that of a companion with the combined mass of smaller SSCs.

3.2.4 Rates of Li Enrichment

We now predict the rate of Li enrichment in giants by the engulfment of SSCs. To do this, we need to use known distributions of planetary periods and masses up to 15 M_J and simulate a population of stars with different properties.

We use the planetary mass distribution of Udry and Santos (2007), which shows a steep rise in the region of low-mass planets. To test if the chosen planetary distribution affects our simulated populations, we also use the distribution of Howard et al. (2010) that extends toward even lower planet masses. However, our results show that the very low mass planets included in this distribution produce negligible changes in the surface Li abundance of stars and do not affect the final Li distribution. We extend the distributions to include BDs with a limiting mass of 15 M_J. The mass distribution alongside the period distribution for the simulated SSCs is in Fig. 3.11.

The period distribution considered for planets comes from Howard et al. (2012), for both smaller planets (2–4 $M_\oplus$) with periods $P < 50$ days and for giant planets (8–32 $M_\oplus$). In this last regime, the period distribution is consistent with that of Cumming et al. (2008) who also include planets with longer periods up to $P = 2000$ days.

Our synthetic populations of stars are separated into different mass regimes, between 1.0 and 2.0 $M_\odot$, with metallicities in the range [Fe/H]=−2.0 to [Fe/H]=0.0.

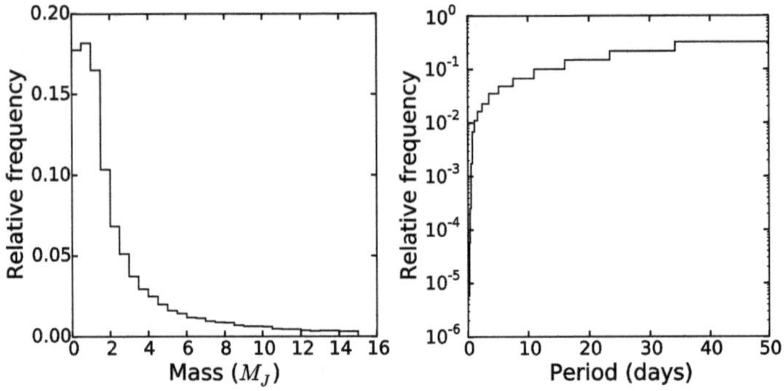

Fig. 3.11 Distribution of planet masses (left panel) and periods (right panel) used to simulate populations of canonical giants and stars accreting SSCs. Credit: Aguilera-Gómez et al. (2016b)

The known Li-rich giants are usually concentrated toward solar metallicity, an effect that may be related to how these objects are found. Thus, our simulated giants follow a similar distribution. Additionally, the simulated stars engulf SSCs of periods and masses taken from the chosen distributions. Given the planetary and stellar properties, we interpolate our grid of Li abundances post-engulfment and obtain the Li abundance after accretion of the specific SSC.

The different stellar mass regimes simulated also contemplate a specific initial Li abundance, where low-mass, old halo stars have $A(Li)_{*,ini} = 2.6$, intermediate star from the Li dip have $A(Li)_{*,ini} \ll 1.0$, and higher-mass stars of 2.0 $M_\odot$ will have an initial Li of $A(Li)_{*,ini} = 3.3$ (corresponding to stars with ages < 2.0 Gyr). Considering the different mass domains also ensures that we take into account the age distribution of stars which implies a specific abundance.

The distribution of A(Li) in simulated populations of 35,000 stars that have engulfed their respective SSCs can be seen in Fig. 3.12. The different mass ranges are indicated at the top of each panel, with the last distribution at the bottom right being the distribution of field giants by Adamów et al. (2014). We compare each of the simulated distributions with a synthetic population of giants with standard stellar evolution. The reported abundance in this figure was evaluated at $\log(g) = 1.5$.

The population of low-mass stars evolving canonically always shows abundances lower than A(Li) = 1.5, as usually reported. However, standard higher-mass stars can have an abundance higher than that fixed limit without accreting any SSC. In other conditions, these stars could be considered as lithium enriched with the usual criteria to distinguish giants, when they are just standard, "normal" stars.

In the low-mass range, there is an increase of giants with Li abundances between A(Li) = 1.5 and A(Li) = 2.2 produced by engulfment that the canonical population does not show. The peak of the Li distribution for giants with engulfment is shifted toward higher abundances, with a decreasing tail that reaches A(Li) = 2.2.

Focusing on the higher-mass stars, both distributions of canonical giants and those engulfing a planet are very similar. Thus, this is a bad range of masses to find Li-enriched giants, which are indistinguishable from the standard population.

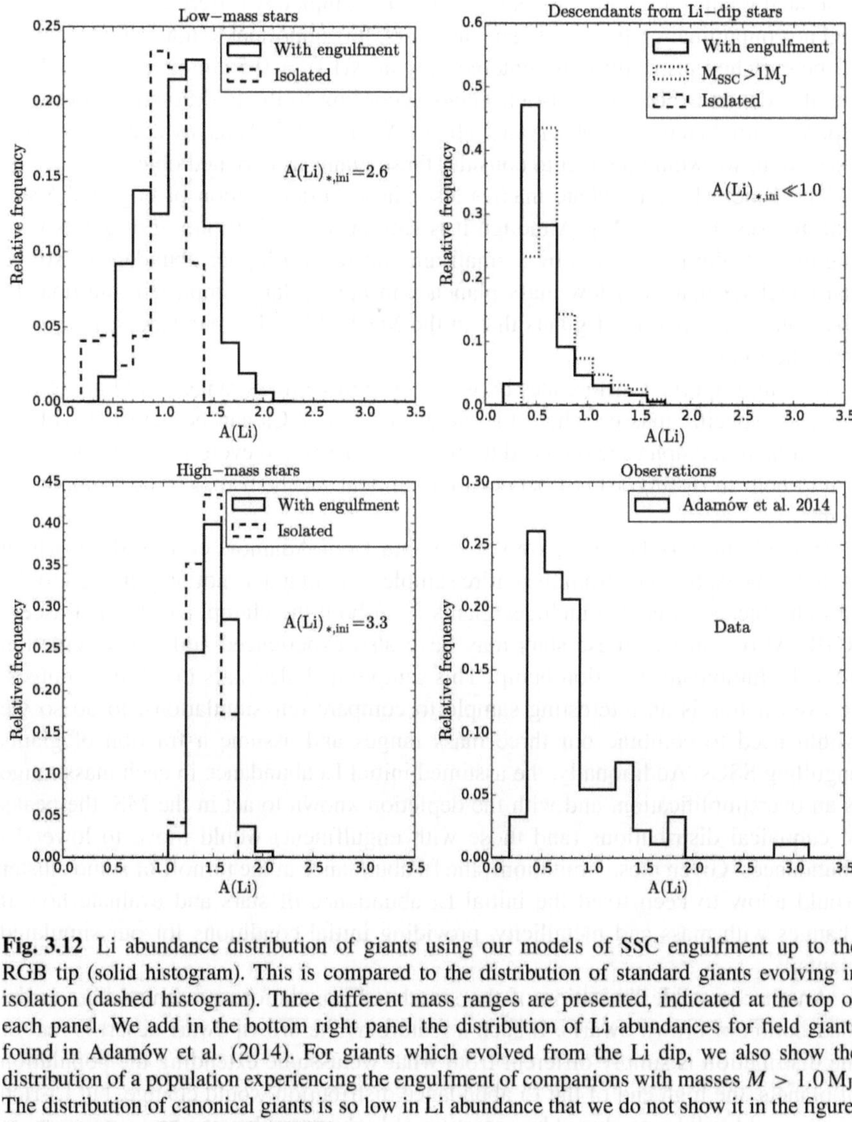

Fig. 3.12 Li abundance distribution of giants using our models of SSC engulfment up to the RGB tip (solid histogram). This is compared to the distribution of standard giants evolving in isolation (dashed histogram). Three different mass ranges are presented, indicated at the top of each panel. We add in the bottom right panel the distribution of Li abundances for field giants found in Adamów et al. (2014). For giants which evolved from the Li dip, we also show the distribution of a population experiencing the engulfment of companions with masses $M > 1.0\,M_J$. The distribution of canonical giants is so low in Li abundance that we do not show it in the figure. Credit: Aguilera-Gómez et al. (2016b)

The intermediate-mass range is a special case where the MS stars have extremely low Li abundances, only measured as upper limits. As such, we cannot see the distribution of canonical giants in Fig. 3.12. As upper limits are not actual measurements, the location of this peak changes depending on the assumed abundance for these giants. However, being the initial abundance so low, it does not affect the abundance post-engulfment. Stars evolving from the Li dip are an interesting sample to study a clean sample of giants being enriched by engulfment, because the distribution of canonical giants and those with engulfment are completely different, and we would find no contamination from isolated giant stars. It is important to notice that the peak of the enriched distribution is centered only at A(Li) $= 0.5$ almost never reaching the abundance expected in Li-rich giants according to the traditional definition to separate enriched objects observationally of A(Li) $= 1.5$. Thus, by considering this fixed limit, we would neglect to consider these giants as enriched objects.

The panel of intermediate masses also shows a distribution of engulfed SSCs with masses $M > 1.0\,M_J$. Although it is similar to the distribution of giants with engulfment, the peak shifts by a small amount toward higher abundances, so we can conclude that even low-mass planets can have a large impact in the final Li abundance distribution of giants that in the MS had low Li abundances, i.e., those from the Li dip.

Finding important differences between these three mass ranges and finding that there are specific masses where the engulfment of SSCs can be detected without contamination emphasize the need to measure masses and evolutionary stages (that would help in distinguishing between enrichment mechanisms) in our samples of giants.

In the bottom right panel, we see the data from Adamów et al. (2014), which includes measurements from the entire sample of field giants not only limited to the Li-rich objects. This also includes giants from both the clump and the first-ascent RGB. Also, some of these stars may have also experienced additional depletion after the luminosity function bump. This sample includes stars in all mass ranges, so even if this is an interesting sample to compare our simulations, to do so we would need to combine our three mass ranges and assume a fraction of giants engulfing SSCs. Additionally, the assumed initial Li abundance in each mass range is an oversimplification, and with the depletion known to act in the MS, the peaks of canonical distributions (and those with engulfment) would move to lower Li abundances. Given these limitations, the Li abundance at the turnoff of some cluster would allow to keep fixed the initial Li abundance of stars and evaluate how it changes with mass and metallicity, providing initial conditions for our simulated giants.

Another possible limitation of the simulated populations presented here is the distribution of brown dwarfs. If such a feature as the brown dwarf desert is real or the distribution is simply different from what we assume extending the population of planets, the high end of the Li abundance distribution would change, but Li-rich giants could still be produced by accretion of high-mass planets.

We then conclude that SSC engulfment can explain stars with high Li abundances, but it can clearly not explain the peak in the observed distribution close to A(Li) $= 3.0$. These giants need a different source of Li or noncanonical mechanism to be explained.

The future of the study of Li-rich giants is in surveys that provide new ways to measure the mass and evolutionary stage of giants, such as *Gaia* parallaxes, asteroseismology with TESS, Kepler, and K2, plus the spectroscopic data to obtain the Li abundance and other important indicators.

An important conclusion from our models is that the definition of what is considered a Li-rich giant in the literature is a limit based on the abundance and dilution in the Sun and corresponds to a fixed value (A(Li) $>$ 1.5), regardless of stellar properties. Using this limit would leave out some giants that have been enriched and cannot be explained by canonical stellar evolution. This limit will also include in the population of enriched giants some objects that are standard, with no other process affecting them, as we see in the Li abundance distribution of high mass stars, where the canonical giants can have A(Li) $=$ 1.7 or higher. In consequence, low- and intermediate-mass giants that engulfed a SSC could be classified as normal giants, and higher-mass canonical stars would be mistaken as Li-rich.

We calculate some rates of Li enrichment by engulfment of SSC considering the traditional definition of A(Li) $>$ 1.5 in the different mass ranges. To do this, we combine the fraction of giants that are Li-rich in our models with engulfment with the fraction of stars that have close-in planets during the MS that could be engulfed in the RGB.

Considering a fraction of 10% and 30% of stars with planets orbiting them in the MS, in the range of low-mass stars, the fraction of enriched giants (A(Li) $=$ 1.5) would be from 1% to 3%, similar to the fraction reported in the literature. With that limit of A(Li) $=$ 1.5, intermediate-mass stars produce a smaller rate of enrichment, with 0.2% to 0.6% of the giants being enriched. This is misleading because every star with a measured abundance in this mass range would be enriched, while every canonical giant only shows upper limits. Any planet engulfed by these stars would produce a visible signal. The importance of this mass range is not the creation of giants with very high abundances but the production of a clean sample of giants where enrichment mechanisms could be studied.

Finally, the rate of enriched giants in the high-mass regime would be from 3.4% to 9.6%, a much higher fraction than for low-mass stars. This result, once again, is produced by the use of the misleading limiting abundance of A(Li) $=$ 1.5, and it is significantly higher because it includes a large number of canonical giants.

3.2.5 SSC Engulfment Models in Perspective

Understanding the trends observed for the canonical giants is the first step in analyzing models of SSC engulfment and noncanonical physics. We present in Fig. 3.13 the Li dilution factors in standard stars going through FDU with different properties, defined as

$$\text{Depletion} = A(\text{Li})_{\text{MS}} - A(\text{Li})_{\text{post FDU}}. \tag{3.4}$$

From this definition, a depletion factor of 1.0 is a reduction of a factor of ten in the Li abundance during FDU.

In principle, the FDU is weakly dependent on mass and metallicity in most cases. However, Fig. 3.13 shows that the higher-metallicity stars of 1.0 $M_\odot$ show a large

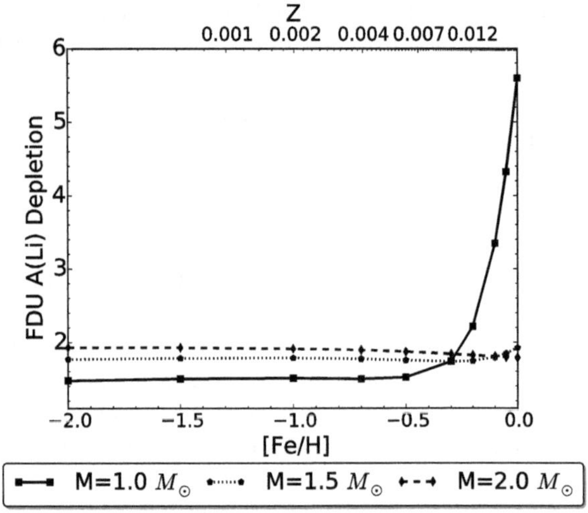

Fig. 3.13 Canonical depletion factors of Li as a function of metallicity during the FDU. This compares the abundance of the star before starting the FDU and after the convective envelope has reached its maximum depth. Low-mass stars with high metallicity combine the effect of dilution and burning to produce higher depletion factors. Credit: Aguilera-Gómez et al. (2016b)

depletion factor that is not only produced by dilution but also due to the burning of Li in convective conditions as the star reaches the RGB.

The large depletion factors of low-mass metal-rich stars and the corresponding low Li abundances that we expect in the RGB make them an interesting sample to study noncanonical physics.

Known Li-rich giants, selected by the threshold A(Li) > 1.5, are compared to our models in Fig. 3.14. The maximum abundance that we predict in giants using our models after the engulfment of different SSCs is depicted as horizontal lines. The vertical solid line indicates the position of the luminosity function bump for a $1.0\,M_\odot$ star of solar metallicity. This is one of the keys to finding the mechanism behind the Li enrichment, so it is important to notice that the position of the RGB bump changes depending on the properties of the star. For all the remaining modeled stars, the RGB bump is always located at lower $\log g$.

From all modeled stars and SSC combinations, the highest Li abundance obtained after SSC accretion in the star is A(Li) $= 2.2$. Giants with abundances higher than that must be explained by other mechanisms entirely or by a combination of external accretion and other processes. Before the RGB bump, SSC engulfment is one of the few mechanisms known to work, and thus Li-rich giants confirmed in this evolutionary phase are likely formed by this mechanism, as long as their abundance is A(Li) ≤ 2.2.

The planetary properties that are more important in the Li enrichment of the star are the mass and metal content of the SSC. The metallicity of the star changes the Li by at most 0.5 dex and the orbital period produces negligible effects on this abundance. As we have seen, the mass of the star is also an important factor in the final Li abundance.

This result, indicating the maximum Li abundance expected from engulfment, agrees with an order of magnitude estimates presented in the literature. The

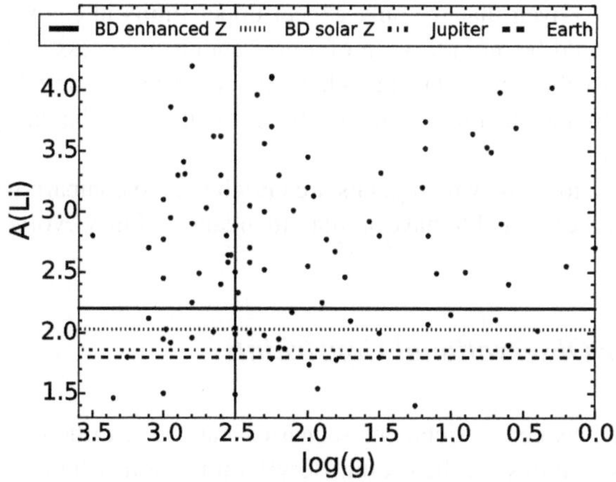

Fig. 3.14 Observations of Li-rich giants compared to the maximum obtained A(Li) from models of engulfment with SSC with four different masses indicated by the horizontal lines. The vertical line shows the location of the luminosity function bump for the 1.0 $M_\odot$ of [Fe/H]=0.0. Other modeled stars have their RGB bumps at lower $\log(g)$ values. Here it is evident that most of the giants with smaller Li abundances could be explained by SSC engulfment, but the giants with higher A(Li) need a different source. The observations of Li-rich giants are obtained from Brown et al. (1989); Luck and Heiter (2007); Ruchti et al. (2011); Kumar et al. (2011); Kirby et al. (2012); Martell and Shetrone (2013); Adamów et al. (2014), and references therein. Credit: Aguilera-Gómez et al. (2016b)

conclusion usually indicates that we must consider realistically the amount of Li that can be provided by the planet and the total enrichment that can be produced by the enrichment event, considering if specific giants can be examples of accretion (Carlberg et al. 2013; D'Orazi et al. 2015; Carlberg et al. 2015). Although we agree with this general conclusion, our work allows to quantify the impact of the different parameters of both planet and star in the Li abundance, improving these general remarks.

Our calculated rates of Li enrichment (Sect. 3.2.4) can be useful to compare with observations previous to the luminosity function bump, where there are no enriched giants produced by other enrichment mechanisms and no extra-mixing affecting stars. However, we would like to emphasize that these rates are obtained with the traditional definition of what is a Li-rich giant, and as such, they can be misleading. Rates for lower-mass stars neglect some of the enriched objects because they have smaller abundances, while in the high-mass range (up to 2.0 $M_\odot$), the used limit selects canonical giants that normally would have higher abundance. This limit also predicts erroneous small rates for giants evolved from the Li dip, as in this mass range, any giant with a measurable Li abundance would be enriched, given the low initial stellar Li.

The sample of giants evolved from the Li dip and before the luminosity function bump is ideal to study the engulfment of planets. It is a clean sample with no other effect on the abundance, where the enriched giants cannot be confused with the canonical giants. Given that we only expect upper limits in canonical giants, any measurement would be an indication of a noncanonical process and it could allow to explore the distribution of planetary characteristics (considering that these planets were engulfed by the star) in a range of masses that cannot be easily probed with radial velocities.

An alternative to know which giants are enriched is to compare with giants with similar properties expected to have similar abundances if they evolve canonically.

3.2.6 *Other Observational Constraints*

Discriminating between enrichment scenarios can be a difficult task, but other observational signatures can be useful. Beryllium is another light element that can be hard to measure, but that has been used to distinguish enrichment mechanisms. We include it in our models for a quick order of magnitude estimate. Assuming that a BD of 15 M_J with meteoritic composition (A(Be)=1.42, Lodders and Fegley (1998)) is ingested by the stars, in the best possible case, an abundance of $A(Be) = 0.7$ post-engulfment is found, with the abundance of the canonical star after the FDU being $A(Be) = 0.6$. This means that the increase in Be in the stellar envelope is very small. Melo et al. (2005) try to measure the Be abundance, finding that all of their giants, even if they have high Li, are extremely depleted in Be. According to their results, a giant with post-FDU abundance of $A(Be) = -5.0$, after accreting a SSC, should have $A(Be) = -0.22$, a signal Melo et al. (2005) indicate should be detectable. If the Be abundance in the canonical giants is as low as indicated in that work, every SSC could increase the Be especially those objects that can also increase the Li abundance, and the lack of Be in Li-rich giants would indicate that they must be produced by a different mechanism. If, on the contrary, stars could have higher Be abundances before the engulfment, the SSC would produce only a small, barely detectable signature on the final Be, being the final abundance very similar to an isolated giant.

A faster rotation has also been associated with Li enrichment. We compare our fraction of Li-enriched giants with the fraction of rapid rotators by Carlberg et al. (2009). Their sample of stars has an average mass of $M = 1.1\,M_{\odot}$, so to compare we use our modeled stars in the low-mass regime. With assumptions on the fraction of stars with planets orbiting them, the fraction of planets that could spin-up the star, and the RGB lifetime, they calculate a fraction of rapidly rotating red giants by engulfment of 0.53% to 0.58%. This fraction is lower than the fraction of enriched giants we find in that mass regime of 1%, but if rapidly rotating giants are able to slow down their rotation while retaining the Li in their envelopes, we could expect a lower fraction of rapid rotators than of Li-rich giants. Some of the assumptions used, such as the fraction of stars that have planets orbiting them in the MS, are somewhat different between works. Additionally, the lower fraction of rapidly rotating stars could be produced if the accretion of some SSC is able to increase the Li abundance without increasing its rotation rate.

Looking for correlations between Li enrichment and other signatures such as rapid rotation (Drake et al. 2002; Carlberg et al. 2012) is fundamental in finding if these giants could be explained by SSC engulfment. However, when working with such samples, stars with abundances higher than A(Li) ~ 2.2 should be treated with caution as they are most likely produced by a different enrichment mechanism.

The reported maximum Li abundance that can be obtained with SSC engulfment has some limitations, as it is only based in the dilution of the SSC content in the stellar envelope. It is also possible that the presence of planets around stars is important not only because they can deposit their Li in the convective envelope but also because accreting mass from a SSC could trigger some kind of extra-mixing mechanism or enhance the mixing rates, producing even higher Li abundances. Some studies in open clusters (Delgado Mena et al. 2016; Carlberg et al. 2016) suggest that RGB stars hosting planets have overall higher Li abundances, but it is unknown how this relates to the presence of SSCs. As such, even when SSC engulfment by itself cannot explain all observations, the presence of SSCs may be relevant for the Li enrichment of giants, as suggested by Adamów et al. (2014), specifically above the RGB bump where the SSC could act as an indirect trigger of mixing.

3.2.7 *Evolution of* ^{6}Li *Abundance*

Just like the more abundant stable isotope of lithium ^{7}Li, lithium-6 (^{6}Li) is very fragile. It is destroyed at even smaller temperatures than ^{7}Li, so consequently, standard depletion models also predict very small ^{6}Li abundances in stellar interiors during the main sequence and very low surface abundances during the RGB phase.

The abundance of ^{6}Li in Li-rich giants has been used to distinguish between replenishment scenarios. The Cameron-Fowler mechanism (plus extra-mixing) only produces ^{7}Li and is unable to produce ^{6}Li. In contrast, brown dwarfs and planets do not burn their ^{6}Li, having a large reservoir of the isotope. Thus, if the Li enrichment is produced by SSC engulfment, the ^{6}Li should increase. In our models of SSC engulfment, we also follow the evolution of ^{6}Li.

The assumptions, point of SSC dissipation, and stellar models are the same here for ^{6}Li as what was described in previous sections for ^{7}Li (see 3.2.1). Some details of the models have to be changed to account for the different isotope:

- The stellar and SSC ^{6}Li is burned through the reaction

$$^6\text{Li} + \text{H} \rightarrow {}^3\text{He} + {}^4\text{He}. \tag{3.5}$$

  For this proton capture reaction, we use the rates from Lamia et al. (2013).
- The considered initial ^{6}Li/^{7}Li ratio for the star is the meteoritic value of ^{6}Li/^{7}Li $= 0.082$ (Chaussidon and Robert 1998).

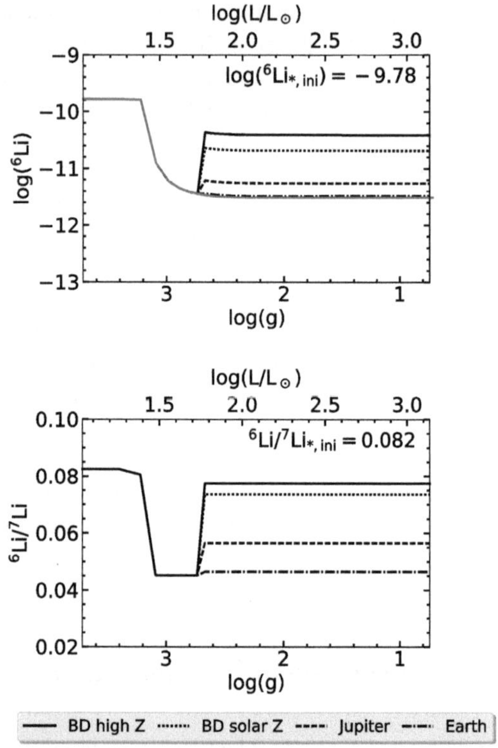

Fig. 3.15 Evolution of surface ^{6}Li abundance (top panel) and lithium isotopic ratio ^{6}Li/^{7}Li (bottom panel) in 1.0 $M_\odot$ stars of [Fe/H]=−2.0, engulfing four different SSCs (each line) at the same timestep, a brown dwarf enhanced in metals, a brown dwarf with solar metallicity, a Jupiter-like planet, and an Earth-like planet. The continuous gray line in the top panel shows the abundance of the star evolving in isolation

- To obtain the SSC ^{6}Li content, we use a similar approach as that for ^{7}Li, using a fixed ratio between ^{6}Li mass fraction and metals equal to the solar system meteoritic value. Thus, SSCs formed have the same $X_{^6\mathrm{Li}}/Z$ but may have a different metal content Z_{SSC} and, consequently, a different mass fraction of ^{6}Li. We use the same Z_{SSC} as described in Sect. 3.2.1.

Our results are strongly dependent on the combination of mass and metallicity of the star.

The SSCs are as presented in section "Changing the Engulfed SSC": an Earth-like rocky planet with $Z = 1.0$, a Jupiter-mass planet with $Z_{\mathrm{SSC}} = 2.5\,Z_\odot$, a brown dwarf of 15 M_J with $Z_{\mathrm{SSC}} = 2.5\,Z_\odot$, and a brown dwarf of the same mass but solar metallicity.

We present first the results for a star of [Fe/H]=−2.0 and masses of $M = 1.0\,M_\odot$, in Fig. 3.15 and $M = 1.5\,M_\odot$ in Fig. 3.16.

Figure 3.15 shows the behavior of ^{6}Li (top panel) and the ratio ^{6}Li/^{7}Li (bottom panel) with $\log g$ and luminosity in a 1.0 $M_\odot$ star after the engulfment of the four different SSCs just described. The gray line in the top panel is the abundance of the star evolving in isolation. A comparison of the abundance of ^{6}Li and that of ^{7}Li shows that their behavior is very similar in this metal-poor star. We see a constant abundance previous to the FDU that then decreases when it is diluted

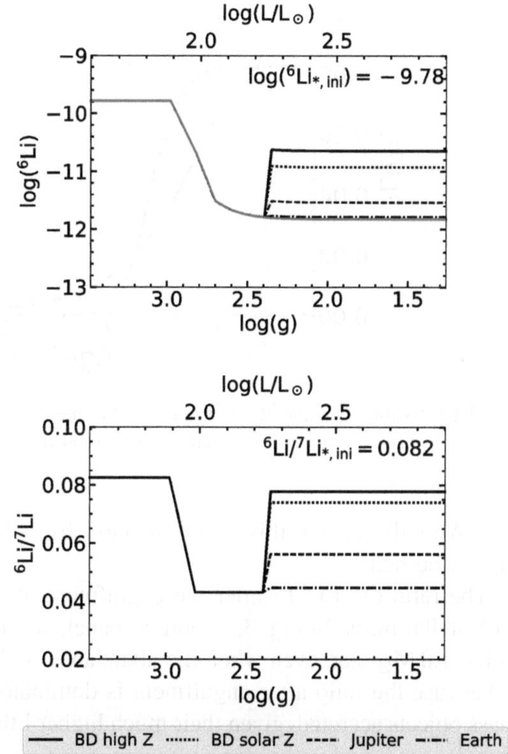

Fig. 3.16 Same as Fig. 3.15 but for a $1.5\,M_{\odot}$ star of $[Fe/H] = -2.0$ after the engulfment of the same 4 SSCs

by the expanding convective envelope. After the engulfment, the abundance has a rapid increase, with almost constant abundance afterward. As in the case of ^{7}Li, the increase in abundance after the engulfment is higher for the brown dwarf with enhanced metal content.

The abundance of ^{6}Li is usually reported as a function of the abundance of ^{7}Li. In the bottom panel of Fig. 3.15, we show the ratio $^6\mathrm{Li}/^7\mathrm{Li}$, constant at meteoritic value before the FDU, constant at a value ~ 0.04 before the engulfment, and constant after accreting the SSC, the value depending on the accreted object. The decrease in the ratio after the FDU is produced because of the different burning rates of ^{6}Li and ^{7}Li. Then, in the layers right below the convective envelope during the MS, more ^{6}Li is burned than ^{7}Li, and when mixing this material during the FDU, ^{6}Li is reduced even more than ^{7}Li, decreasing the ratio.

In these low metallicity stars, due to the combination of densities and temperatures found in the bottom of the convective layer after the FDU, the ^{6}Li is not burned significantly during the RGB. It is in these stars where the $^6\mathrm{Li}/^7\mathrm{Li}$ ratio can reach a maximum value after engulfment, almost recovering the meteoritic value $^6\mathrm{Li}/^7\mathrm{Li} = 0.078$ after the engulfment of a $15\,M_J$ brown dwarf enhanced in metals. An Earth-like object almost does not increase the ^{6}Li in comparison with an isolated

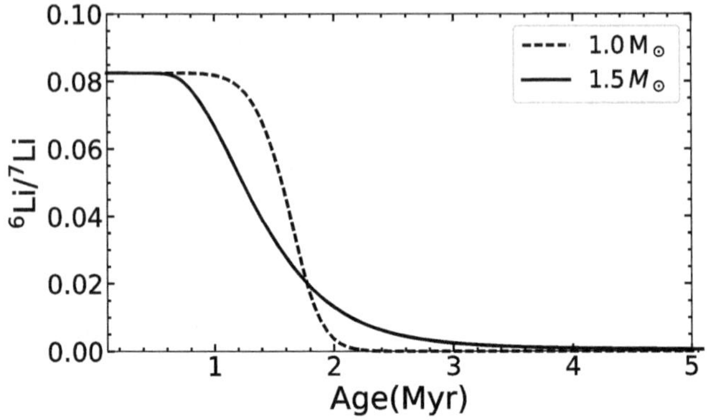

Fig. 3.17 Evolution of the ^{6}Li/^{7}Li ratio in solar metallicity stars of 1.0 and 1.5 $M_\odot$. These show that the ^{6}Li is destroyed and completely depleted during the pre-MS

star. We will see that this is not the case for solar metallicity stars, where ^{6}Li is rapidly burned.

The ratio of ^{6}Li/^{7}Li after the engulfment of the SSC almost does not change with stellar mass. In Fig. 3.16 bottom panel, we see similar ratios in a 1.5 $M_\odot$ than in the 1.0 $M_\odot$ star, even when the abundance of ^{6}Li is different in both stars. This is because the ratio after engulfment is dominated by that of the SSC for higher-mass objects accreted, given their much higher lithium abundances compared to the abundances found in the star.

Results for stars of [Fe/H]= 0.0, i.e., solar metallicity, and masses of M = 1.0 $M_\odot$ and 1.5 $M_\odot$ are very different. Canonical stellar models predict a rapid burning of ^{6}Li during the pre-MS, in the first Myr of life for the stars (Fig. 3.17). The abundance of ^{7}Li also decreases during the pre-MS in these stars, but a lot of it remains. In the figure, the abundance starts to decrease when the star contracts enough to reach temperatures for Li burning.

As ^{6}Li is already completely destroyed during the pre-MS, RGB stars are expected to have no ^{6}Li. The engulfment of a companion can greatly increase the abundance and the ratio of ^{6}Li/^{7}Li, even reaching near to meteoritic values, but for these stars, the ^{6}Li is burned very rapidly after the engulfment in the convective layer, in timescales of $\sim$ 100 Myr. The exact maximum abundance depends on the time of engulfment, as was the case for ^{7}Li. After burning the ^{6}Li deposited by the SSC, the 1.0 $M_\odot$ star completely destroys it, but the 1.5 $M_\odot$ star keeps an extremely low abundance, very close to zero in the envelope. In contrast, ^{7}Li can be preserved for the whole RGB (without considering extra-mixing processes that could decrease the abundance). We have also found the limiting metallicity for which the ^{6}Li is burned in the convective envelope, meaning that after the engulfment, this isotope is not preserved. For 1.0 $M_\odot$ stars, the limit is between [Fe/H] = −0.5 and −0.3, meaning that everything more metal-poor preserves some of its ^{6}Li, although not all of it, while for the 1.5 $M_\odot$ stars, this limit moves to higher metallicities, closer to [Fe/H] = −0.1.

This lead us to important conclusions regarding the presence of ^{6}Li used as a method to distinguish between mechanisms. The absence of the element does not immediately discard the external mechanisms, such as the engulfment of a SSC. For higher-metallicity stars, the element is rapidly burned, so its absence could indicate that the engulfment process is not recent, but not that there was no engulfment at all, similar to what Drake et al. (2002) concluded. On the other hand, our results on ^{6}Li coupled with those for ^{7}Li show that any star with an abundance of A(^{7}Li) $>$ 2.2 cannot be explained by SSC engulfment, regardless of the presence or absence of ^{6}Li. Thus, finding extremely enriched stars with ^{7}Li that do not have any ^{6}Li (e.g., Monaco et al. 2014), even if they were metal-poor, does not say anything about the SSC engulfment scenario.

On the other hand, the detection of ^{6}Li in a Li-rich giant strongly argues in favor of an external scenario, as the high abundance of both isotopes is unusual and should be explained separately if SSC engulfment is not accepted as an explanation. Other mechanisms, such as stellar flares, can produce ^{6}Li. The high abundance of this isotope in the solar wind based on measurements of the lunar soil (Chaussidon and Robert 1999) indicates that it is possible that the Sun is producing ^{6}Li via flares. If part of the ^{6}Li is retained by the photosphere of the star instead of being transported outward by the flares, the abundance would still be too low to be detected in measurements of the photosphere of the Sun, where no ^{6}Li is found. Considering the larger convective envelope of a giant, any ^{6}Li produced by this mechanism would be diluted, and it could not be observed on the surface of the star. But, even if solar flares could increase the ^{6}Li, another process should be invoked to explain the high abundance of ^{7}Li (see Mott et al. 2017, for a discussion on a Li-rich giant with ^{6}Li).

From a purely observational point of view, the isotope ^{6}Li is particularly hard to measure in stellar spectra, as it manifests only as a subtle asymmetry of the ^{7}Li doublet. Only with very high resolution and signal-to-noise spectra can this feature be measured. We need to consider this when trying to measure ^{6}Li for Li-rich giants. Also, the larger the ^{7}Li abundance, the easier it is to detect the asymmetry produced by ^{6}Li for the same quality spectra. As we have found a limit to the ^{7}Li in Li-rich giants produced by engulfment (process that can also increase the ^{6}Li), a detection of ^{6}Li would be especially interesting in Li-rich giants with A(^{7}Li) $<$ 2.2, for which the detection of ^{6}Li requires even better quality spectra than to detect the ^{7}Li.

In conclusion, the absence of ^{6}Li may not be a good diagnostic against SSC engulfment and should be treated with caution when used in favor of one or other scenario of enrichment.

3.3 Summary

In this chapter, we studied the external scenarios that could produce an enhancement in Li abundance in giants.

We have studied RV variability of Li-rich giants in the literature. Most of them are consistent with having no stellar mass companions in the AGB phase or white dwarf companions that could evidence a past event of mass transfer. As such, these Li-rich giants were never able to accrete material from another star. This leaves as a source of pollution the substellar mass companions, such as brown dwarfs and planets.

We modeled the engulfment of SSCs by stars, exploring a wide parameter range, changing stellar masses and metallicities, and planetary properties such as composition, period, and mass. With these calculations, we can obtain a maximum lithium content expected in stars during the RGB phase after the accretion of SSCs. This is especially interesting before the luminosity function bump, where extra-mixing does not seem to be efficient to produce Li-enriched giants, and there is no extra-mixing modifying the surface composition of giants.

To deposit the fresh Li from the SSC inside the star, we first calculate the point inside the star where the planet/BD would dissolve. For planetary masses higher than $M_{\rm SSC} > 15\,\rm M_J$, the dissolution is produced in the radiative interior, and in consequence, the SSC does not produce a visible signature in the stellar surface. Smaller SSCs can increase the surface Li of the star reaching values that are at most of A(Li) $= 2.2$. Assumptions were considered in our modeling process, always pointing to produce a trustworthy upper limit, so giants with abundances much higher than this limit may need either a different noncanonical mechanism to be explained or a combination of SSC engulfment and other methods, such as mixing.

Earth-like planets can produce only a very small signature in the Li abundance of the star. The largest increase found is produced by a brown dwarf enhanced in metal content. For this last case, where the Li of the SSC is abundant, the initial Li abundance of the giant almost does not change the Li post-engulfment. However, when a smaller object with lower Li content is engulfed, the initial Li of the star can produce important variations on the final Li after the accretion event.

Separating into different mass ranges, so that we can consider the possible differences in surface Li of stars, reveals interesting results. For low-mass stars ($1.0 \leq M < 1.3\,\rm M_\odot$) with an initial abundance of $\rm A(Li)_{*,ini} = 2.6$, we calculate that SSC with masses $M > 6.7\,M_{\rm J}$ produce a signature in Li that can be detected and would not be confused with canonical giants. We calculate the fraction of Li-rich giants with abundances A(Li) $\gtrsim 1.5$. This corresponds to a 10% of the simulated population that always accretes one SSC, which translates into a 1% to 3.% of the entire population of giants being enriched in Li.

Stars in the intermediate-mass ranges are those evolved from the Li dip ($M \sim 1.3\rm M_\odot$). We consider for these an initial abundance of $\rm A(Li)_{*,ini} \ll 1.0$, and, given the extremely low abundance of canonical giants, all planets would increase considerably the surface Li abundance of the star. Thus, this would be an interesting sample to study SSC engulfment. Giants evolved from the Li dip with abundances A(Li) $\gtrsim 1.5$ after engulfment are only a 2% if the entire synthetic population accretes SSCs. Assuming only a fraction of the population of giants could engulf planets, we find rates of enrichment of 0.2% to 0.6% a misleading fraction considering that each detection in this mass range would point to a previous accretion event.

Higher-mass stars (up to $2.0\,M_\odot$) have initial abundances of $A(Li)_{*,ini} = 3.3$, which leads to higher abundances post-FDU. Identifying the signal from an engulfed SSC in this mass range would be much harder because the population of canonical giants has a similar Li distribution than that with engulfment. A minimum mass of $M_{SSC} > 6.0\,M_J$ must be consumed by the star. Considering the criteria $A(Li) > 1.5$ of selection of Li-rich giants leads to a misleading large fraction of enriched giants. A 32% of stars that have engulfed a planet are classified as Li-enriched, corresponding to a 3.4% to 9.6% of the total population. This mass domain is not the best to identify enriched giants because the distribution with and without engulfment are very similar.

All these different fractions suggest that the currently used traditional definition of Li-rich giants, with the threshold $A(Li) = 1.5$ can be misleading. Some of the low-mass and intermediate-mass stars that have accreted SSCs would not be considered enriched by that criterion, losing part of the sample that is being produced by engulfment of planets or BD. On the contrary, some of the canonical higher-mass stars would be considered as enriched even if they are not subjected to any nonstandard physics. In consequence, it is really important to study samples that have well-measured masses and evolutionary stages.

Among all the explored parameters, the period of the planet does not greatly affect the Li abundance post-engulfment, being the most important planetary characteristics the mass and composition. Regarding the host star, changes in the mass of the star can produce differences of 0.5 dex in Li after engulfment, without considering realistic mass domains and their corresponding initial Li abundances. When taking this into account, the mass of the star is extremely important not only on the final Li abundance but also on the correct identification of enriched sources and their detectability. In the higher-mass domain, the metallicity does not produce large effects. The [Fe/H] seems to be more relevant in low-mass stars given its effects on the size of the convective envelope after FDU.

The SSC engulfment mechanism is promising to explain Li-enriched giants in specific domains of parameter space where extra-mixing cannot act. Based on current evidence, it should be especially important in metal-rich stars and giants located before the luminosity function bump but after the FDU. The ingestion of a SSC can increase the Li abundance of the giant up to $A(Li) = 2.2$, so if giants in these domains are found to have higher abundances, the dissolution of SSC content in the star alone cannot explain their high Li content. These cases are especially interesting. We see then how the precise evolution stage of the giants is necessary to confirm the enrichment mechanism. It is even possible that some of the Li-enriched giants are located in the clump, so another mechanism could act toward the end of the RGB phase or the He-flash to explain the high Li in those giants.

Ideal locations to study the engulfment of SSC are before the luminosity function bump but also on stars that have evolved from the Li dip, given their extremely low initial Li abundances and the clean sample of enriched objects that we would find in this mass domain.

^{6}Li has been invoked as a possible signature of engulfment in giants. We have also included this isotope in our models. We find that, for low-metallicity stars, the behavior of ^{6}Li is very similar to that of ^{7}Li, increasing right after the engulfment and leaving a visible signature at least until the RGB tip. In contrast, solar metallicity giants do not experience this increase, because the temperatures and densities in the bottom of the convective envelope are high enough to trigger the rapid burning of this isotope. In timescales of 100 Myr, ^{6}Li can be completely destroyed.

Thus, the lack of ^{6}Li does not necessarily imply that the giant has not experienced SSC engulfment. Also, ^{6}Li is easier to measure in giants with very high A(^{7}Li), which cannot be explained by SSC engulfment anyway. Not finding this isotope in stars with A(Li) > 2.2 does not say anything about the enrichment mechanism. We also notice that depending on the atmospheric parameters of the giant, it can be hard to detect ^{6}Li in Li-rich giants that could have been produced by SSC engulfment. On the other hand, finding ^{6}Li in a giant that is enriched in ^{7}Li is a strong indication of external pollution, as the Cameron-Fowler mechanism is only able to produce ^{7}Li.

Our models take into consideration some assumptions. They could improve if we add tidal evolution, extra-mixing, and feedback on the stellar structure. Although we have assumed some initial conditions for our giants based on empirical data, we could improve this by calibrating our models with the turnoff abundance of some clusters, with known masses and metallicities, to generate more realistic populations. However, the numerical tools that we have developed and the conclusions regarding the importance of mass in separating giants into canonical and enriched are the base of future work, where we can add extra-mixing and other parameters that could complicate the interpretation of results but also generate a much more realistic picture of stars in some domains of stellar evolution (i.e., after the RGB bump).

References

M. Adamów, A. Niedzielski, E. Villaver, G. Nowak, A. Wolszczan, ApJL **754**, L15 (2012). http://doi.org/10.1088/2041-8205/754/1/L15

M. Adamów, A. Niedzielski, E. Villaver, A. Wolszczan, G. Nowak, A&A **569**, A55 (2014). http://doi.org/10.1051/0004-6361/201423400

E.G. Adelberger, A. García, R.G.H. Robertson, et al., Rev. Modern Phys. **83**, 195 (2011). http://doi.org/10.1103/RevModPhys.83.195

C. Aguilera-Gómez, J. Chanamé, M.H. Pinsonneault, J.K. Carlberg, ApJL **833**, L24 (2016). http://doi.org/10.3847/2041-8213/833/2/L24

C. Aguilera-Gómez, I. Ramírez, J. Chanamé, A&A **614**, A55 (2018). http://doi.org/10.1051/0004-6361/201732209

J.B. Alexander, The Observatory **87**, 238 (1967)

I. Baraffe, G. Chabrier, T.S. Barman, F. Allard, P.H. Hauschildt, A&A **402**, 701 (2003). http://doi.org/10.1051/0004-6361:20030252

J.A. Brown, C. Sneden, D.L. Lambert, E. Dutchover, Jr., ApJS **71**, 293 (1989). http://doi.org/10.1086/191375

J.K. Carlberg, K. Cunha, V.V. Smith, S.R. Majewski, ApJ **757**, 109 (2012). http://doi.org/10.1088/0004-637X/757/2/109

J.K. Carlberg, K. Cunha, V.V. Smith, S.R. Majewski, Astronomische Nachrichten **334**, 120 (2013). http://doi.org/10.1002/asna.201211757

J.K. Carlberg, S.R. Majewski, P. Arras, ApJ **700**, 832 (2009). http://doi.org/10.1088/0004-637X/700/1/832

J.K. Carlberg, V.V. Smith, K. Cunha, K.G. Carpenter, ApJ **818**, 25 (2016). http://doi.org/10.3847/0004-637X/818/1/25

J.K. Carlberg, V.V. Smith, K. Cunha, et al., ApJ **802**, 7 (2015). http://doi.org/10.1088/0004-637X/802/1/7

B.W. Carney, D.W. Latham, R.P. Stefanik, J.B. Laird, J.A. Morse, AJ **125**, 293 (2003). http://doi.org/10.1086/345386

G. Chabrier, I. Baraffe, B. Plez, ApJL **459**, L91 (1996). http://doi.org/10.1086/309951

M. Chaussidon, F. Robert, Earth Planet. Sci. Lett. **164**, 577 (1998). http://doi.org/10.1016/S0012-821X(98)00250-7

M. Chaussidon, F. Robert, Nature **402**, 270 (1999). http://doi.org/10.1038/46235

J.P. Cox, *Principles of stellar structure - Vol.1: Physical principles; Vol.2: Applications to stars* (1968)

A. Cumming, R.P. Butler, G.W. Marcy, S.S. Vogt, J.T. Wright, D.A. Fischer, PASP **120**, 531 (2008). http://doi.org/10.1086/588487

R.H. Cyburt, B.D. Fields, K.A. Olive, Phys. Lett. B **567**, 227 (2003). http://doi.org/10.1016/j.physletb.2003.06.026

J.R. De Medeiros, J.D. do Nascimento, Jr., S. Sankarankutty, J.M. Costa, M.R.G. Maia, A&A **363**, 239 (2000)

J.R. de Medeiros, M. Mayor, A&AS **139**, 433 (1999). http://doi.org/10.1051/aas:1999401

J.R. de Medeiros, C.H.F. Melo, M. Mayor, A&A **309**, 465 (1996)

J.R. De Medeiros, S. Udry, G. Burki, M. Mayor, A&A **395**, 97 (2002). http://doi.org/10.1051/0004-6361:20021214

E. Delgado Mena, S. Bertrán de Lis, V.Z. Adibekyan, et al., A&A **576**, A69 (2015). http://doi.org/10.1051/0004-6361/201425433

E. Delgado Mena, M. Tsantaki, S.G. Sousa, et al., A&A **587**, A66 (2016). http://doi.org/10.1051/0004-6361/201527196

P. Demarque, D.B. Guenther, L.H. Li, A. Mazumdar, C.W. Straka, Ap&SS **316**, 31 (2008). http://doi.org/10.1007/s10509-007-9698-y

V. Dobrovolskas, A. Kučinskas, P. Bonifacio, et al., A&A **565**, A121 (2014). http://doi.org/10.1051/0004-6361/201322868

V. D'Orazi, R.G. Gratton, G.C. Angelou, et al., ApJL **801**, L32 (2015). http://doi.org/10.1088/2041-8205/801/2/L32

N.A. Drake, R. de la Reza, L. da Silva, D.L. Lambert, AJ **123**, 2703 (2002). http://doi.org/10.1086/339968

J.W. Ferguson, D.R. Alexander, F. Allard, et al., ApJ **623**, 585 (2005). http://doi.org/10.1086/428642

J.J. Fortney, M.S. Marley, J.W. Barnes, ApJ **659**, 1661 (2007). http://doi.org/10.1086/512120

M. Gim, J.E. Hesser, R.D. McClure, P.B. Stetson, PASP **110**, 1172 (1998). http://doi.org/10.1086/316241

N. Grevesse, A.J. Sauval, SSRv **85**, 161 (1998). http://doi.org/10.1023/A:1005161325181

A.W. Howard, G.W. Marcy, J.A. Johnson, et al., Science **330**, 653 (2010). http://doi.org/10.1126/science.1194854

A.W. Howard, G.W. Marcy, S.T. Bryson, et al., ApJS **201**, 15 (2012). http://doi.org/10.1088/0067-0049/201/2/15

E.N. Kirby, X. Fu, P. Guhathakurta, L. Deng, ApJL **752**, L16 (2012). http://doi.org/10.1088/2041-8205/752/1/L16

Y.B. Kumar, B.E. Reddy, D.L. Lambert, ApJL **730**, L12 (2011). http://doi.org/10.1088/2041-8205/730/1/L12

M. Kunitomo, M. Ikoma, B. Sato, Y. Katsuta, S. Ida, ApJ **737**, 66 (2011). http://doi.org/10.1088/0004-637X/737/2/66
R.L. Kurucz, ApJS **40**, 1 (1979). http://doi.org/10.1086/190589
L. Lamia, C. Spitaleri, M. La Cognata, S. Palmerini, R.G. Pizzone, A&A **541**, A158 (2012). http://doi.org/10.1051/0004-6361/201219014
L. Lamia, C. Spitaleri, R.G. Pizzone, et al., ApJ **768**, 65 (2013). http://doi.org/10.1088/0004-637X/768/1/65
K. Lodders, B. Fegley, *The Planetary scientist's companion / Katharina Lodders, Bruce Fegley.* (1998)
R.E. Luck, U. Heiter, AJ **133**, 2464 (2007). http://doi.org/10.1086/513194
S.L. Martell, M.D. Shetrone, MNRAS **430**, 611 (2013). http://doi.org/10.1093/mnras/sts661
A. Massarotti, D.W. Latham, R.P. Stefanik, J. Fogel, AJ **135**, 209 (2008). http://doi.org/10.1088/0004-6256/135/1/209
C.H.F. Melo, P. de Laverny, N.C. Santos, et al., A&A **439**, 227 (2005). http://doi.org/10.1051/0004-6361:20041805
C. Mendoza, M.J. Seaton, P. Buerger, et al., MNRAS **378**, 1031 (2007). http://doi.org/10.1111/j.1365-2966.2007.11837.x
L. Monaco, H.M.J. Boffin, P. Bonifacio, et al., A&A **564**, L6 (2014). http://doi.org/10.1051/0004-6361/201323348
A. Mott, M. Steffen, E. Caffau, F. Spada, K.G. Strassmeier, ArXiv e-prints: 1704.06460 (2017)
G. Nelemans, T.M. Tauris, A&A **335**, L85 (1998)
N. Nettelmann, A. Becker, B. Holst, R. Redmer, ApJ **750**, 52 (2012). http://doi.org/10.1088/0004-637X/750/1/52
H. Palme, SSRv **92**, 237 (2000). http://doi.org/10.1023/A:1005247329412
M.H. Pinsonneault, C.P. Deliyannis, P. Demarque, ApJS **78**, 179 (1992). http://doi.org/10.1086/191624
M.H. Pinsonneault, S.D. Kawaler, P. Demarque, ApJS **74**, 501 (1990). http://doi.org/10.1086/191507
M.H. Pinsonneault, S.D. Kawaler, S. Sofia, P. Demarque, ApJ **338**, 424 (1989). http://doi.org/10.1086/167210
G. Privitera, G. Meynet, P. Eggenberger, A.A. Vidotto, E. Villaver, M. Bianda, A&A **593**, A128 (2016). http://doi.org/10.1051/0004-6361/201628758
I. Ramírez, J.R. Fish, D.L. Lambert, C. Allende Prieto, ApJ **756**, 46 (2012). http://doi.org/10.1088/0004-637X/756/1/46
R. Rebolo, E.L. Martin, A. Magazzu, ApJL **389**, L83 (1992). http://doi.org/10.1086/186354
L.M. Rebull, J.K. Carlberg, J.C. Gibbs, et al., AJ **150**, 123 (2015). http://doi.org/10.1088/0004-6256/150/4/123
F.J. Rogers, A. Nayfonov, ApJ **576**, 1064 (2002). http://doi.org/10.1086/341894
G.R. Ruchti, J.P. Fulbright, R.F.G. Wyse, et al., ApJ **743**, 107 (2011). http://doi.org/10.1088/0004-637X/743/2/107
I.J. Sackmann, A.I. Boothroyd, ApJL **392**, L71 (1992). http://doi.org/10.1086/186428
I.J. Sackmann, A.I. Boothroyd, ApJ **510**, 217 (1999). http://doi.org/10.1086/306545
E.E. Salpeter, Aust. J. Phys. **7**, 373 (1954). http://doi.org/10.1071/PH540373
E. Sandquist, R.E. Taam, D.N.C. Lin, A. Burkert, ApJL **506**, L65 (1998). http://doi.org/10.1086/311633
L. Siess, M. Livio, MNRAS **304**, 925 (1999). http://doi.org/10.1046/j.1365-8711.1999.02376.x
L. Siess, M. Livio, MNRAS **308**, 1133 (1999). http://doi.org/10.1046/j.1365-8711.1999.02784.x
N. Soker, O. Regev, M. Livio, M.M. Shara, ApJ **318**, 760 (1987). http://doi.org/10.1086/165409
J.E. Staff, O. De Marco, P. Wood, P. Galaviz, J.C. Passy, MNRAS **458**, 832 (2016). http://doi.org/10.1093/mnras/stw331
H.C. Thomas, ARA&A **15**, 127 (1977). http://doi.org/10.1146/annurev.aa.15.090177.001015
S. Udry, N.C. Santos, ARA&A **45**, 397 (2007). http://doi.org/10.1146/annurev.astro.45.051806.110529

J.L. van Saders, M.H. Pinsonneault, ApJ **746**, 16 (2012). http://doi.org/10.1088/0004-637X/746/1/16
P. Ventura, F. D'Antona, MNRAS **402**, L72 (2010). http://doi.org/10.1111/j.1745-3933.2010.00805.x
E. Villaver, M. Livio, ApJL **705**, L81 (2009). http://doi.org/10.1088/0004-637X/705/1/L81
E. Villaver, M. Livio, A.J. Mustill, L. Siess, ApJ **794**, 3 (2014). http://doi.org/10.1088/0004-637X/794/1/3
L. Zeng, D. Sasselov, PASP **125**, 227 (2013). http://doi.org/10.1086/669163

Part II
Observational Applications

Our models of planet and brown dwarf engulfment allow predictions of a maximum abundance for the enriched giants and the location on the HR-diagram where we can find Li-unusual giants created by each of the mechanisms. This, in turn, is a base to produce new observational tests that focus on finding these signatures and constrain the enrichment mechanism in each case.

One of the added difficulties of this problem after our modeling is to find the giants that are truly unusual. Before our work, it was enough to find giant with $A(Li) > 1.5$ to define objects that needed an additional explanation. It is common to find serendipitous discoveries of Li-rich giants in samples not designed to study the specific problem of Li enrichment. This fixed limit to what is considered Li-rich has proven to be useful to find very enriched objects by searching for candidates in low-resolution spectra with strong Li lines that should be detected in giants with high Li abundances only, for later high-resolution spectroscopy follow-up to measure the exact abundance.

In conclusion, this approach of looking for giants with $A(Li) > 1.5$ and defining these as Li-rich is useful to find the most Li-enriched objects and it is more cost-efficient. But, we suggest that a different approach is needed if we want to constrain the enrichment mechanism acting on giants and test the predictions we have developed with our models. To distinguish between enrichment mechanisms, there are some ideal samples that can be used.

In particular, we have seen that the knowledge of a stars' mass is critical in determining the abundance it should have without any additional processes acting on it. This allows identification of outliers, i.e., enriched giants, that require non-canonical physics to be explained and to know the maximum Li abundance that we can expect in giants produced by each of the enrichment scenarios.

The inhibition of mixing in the red giant branch before the luminosity function bump turns this location on the HR-diagram into the perfect place to test external models that are the only ones that could produce an enrichment in this phase. The same is the case of the most metal-rich populations in the Galaxy, where mixing is also supposed to be inefficient.

Another sample that can be very interesting to study any type of enrichment mechanisms, either internal or external, are giants that have evolved from the Li dip. As these stars already have very low Li surface abundances in the MS, once they reach the giant phase they will have even more depleted exteriors. As such, any signal produced by an enrichment mechanism could be easily identified.

In this part of the thesis, I will present three different samples that are interesting to study the Li-enrichment mechanisms. The first is the open cluster Trumpler 20 (Chap. 4). Cluster giants have the advantage over field giants in that they have broadly the same mass, so any star with a higher Li content than other giants in a specific evolutionary phase (i.e., at a given $\log g$ or luminosity) is probably abnormal. The second population is a sample of field giants for which masses can be obtained by using their Hipparcos parallaxes. Other advantages of this sample, its characterization, and measured Li abundances can be found in Chap. 5. The last sample is different, in that the stars are not giants, but dwarfs and subgiants. In particular, we study an interesting group of field stars that have depleted surface abundances, the low Li abundance stars in the region known as the *lithium desert*. We conclude that these might be subgiants that have evolved from the Li dip. As such, they are the progenitors of the Li-depleted giants that are interesting to study the enrichment mechanism. These stars would mark the baseline of Li abundance of stars without additional noncanonical processes acting during the RGB phase, that have already depleted their Li abundances during the MS. In Chap. 6 we show the sample, the detailed analysis of the depleted stars and how we identify that these have evolved from the Li dip.

Chapter 4
Trumpler 20

The evolutionary stage of giants is key in finding the enrichment mechanism, where we can distinguish those that have already finished their FDU, those that are located before the RGB bump, giants after the luminosity function bump, and finally, giants that are now in the clump, steadily burning He in their core.

In the field, masses and evolutionary stages of giants have been historically hard to measure, especially before asteroseismology missions such as Kepler, K2, and TESS and before *Gaia* provided parallaxes for an enormous amount of stars in the sky. Still nowadays, not all of the stars, and in particular not all Li-rich giants, have these data.

Open clusters can help us with this particular difficulty, given that their giants have similar masses. Assuming that cluster members have the same original composition, distance, and age, changes in the surface abundance of stars must be produced by nonstandard processes acting inside the star or by external pollution. In that sense, open clusters are ideal to study the abundance anomalies, given that globular clusters show evidence of multiple stellar populations, a fact that gives the advantage to open clusters where the signatures related to stellar evolution are not contaminated.

Smiljanic et al. (2016) (S16 hereafter) have measured the Li abundance of giants in one of these open clusters, Trumpler 20, in the context of the *Gaia*-ESO collaboration (Randich et al. 2013). They find two giants with higher Li abundance than the rest of the cluster, making this an interesting cluster to study Li enrichment. Trumpler 20 has an age of $\sim$ 1.6 Gyr and [Fe/H]=+0.17. Forty-one stars are observed by the *Gaia*-ESO survey, thought to be likely members by measurement of radial velocities (Donati et al. 2014). Additionally, these stars have information on their carbon and nitrogen, measured by Tautvaišienė et al. (2015), and atmospheric parameters and Li abundance with corrections by nonlocal

This chapter was published in Aguilera-Gómez et al. (2016a).

C. Aguilera-Gómez, *Explaining Lithium Enriched Red Giant Branch Stars*, Springer Theses, https://doi.org/10.1007/978-3-030-02583-0_4

thermodynamic equilibrium by S16. We will use all these combined data in this section.

From the 41 stars likely members of Trumpler 20, S16 report Li for 40 giants and 1 subgiant. Out of the 40 giants, only 4 have detections, while all others have upper limits. Two of these detections have abundances that would normally classify them as Li-rich, $A(Li)_{non-LTE} = 1.54 \pm 0.21$, and $A(Li)_{non-LTE} = 1.60 \pm 0.21$.

S16 have a particular explanation for the Li abundance pattern of the cluster, suggesting that all the stars of the cluster have experienced severe post-MS Li depletion except for the two giants with higher Li, which have inhibited extra-mixing. We suggest an alternative explanation to the pattern, where the two detections with high Li abundance have been enriched in Li, possibly by accretion of a substellar mass companion, and the other stars in the cluster have low Li abundances that reflect their depletion due to first dredge-up, considering they went through main sequence depletion. The remaining giants could have also experienced extra-mixing during the RGB, something that we cannot confirm because they have only Li upper limits, but this is not required to explain the abundances of the large majority of cluster stars. These are two different interpretations of the same data, which we will detail in the following sections. To do this, we first summarize the conclusions of S16 that are later contrasted with our models.

4.1 The Possibility of Extra-Mixing in the Trumpler 20 Giants

After presenting their Li abundance measurements, S16 compare with theory. In particular, they compare their abundances with predictions based on standard evolutionary theory and MS Li solar meteoritic abundance. With that in mind, the giants with high Li align with the predictions, while all the other stars in the clusters are located well below models in Li. Combining the position of these stars in the HR diagram relative to isochrones, and the measurements of carbon and nitrogen and their expected values before and after the FDU, S16 claim that all stars with low luminosities in the sample are core-He burning stars. In consequence, the high Li abundance of two giants indicates that they have preserved their Li, while all other giants experienced mixing on the RGB before reaching their core-He burning phase and possibly also went through additional MS mixing.

Instead, our interpretation of the evolutionary stages and Li abundance pattern of the stars is the following: There is a combination of both first ascent (with inert He cores) and red clump giants, burning steadily He in the core. The carbon and nitrogen measurements of the data can be reproduced without the need to invoke mixing. If canonical evolutionary models with standard FDU are first calibrated to the Li abundance in the turnoff of similar clusters (and thus assuming some MS depletion), instead of assuming a solar meteoritic value, the Li upper limits are broadly consistent with models. A natural approach would be to use the abundance

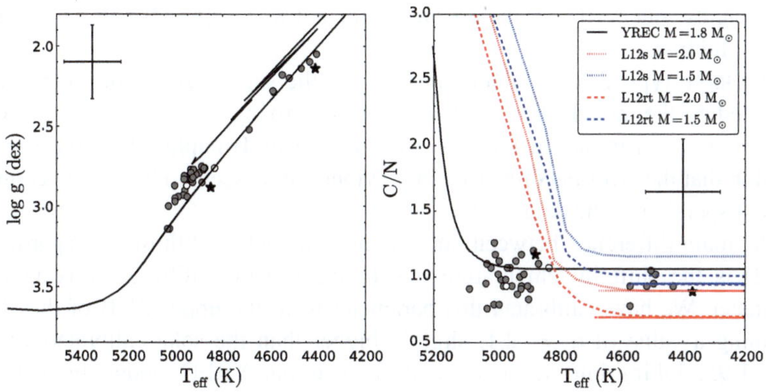

Fig. 4.1 Left panel: HR diagram of Trumpler 20 giants. Star symbols are the two giants with higher Li abundances, filled circles are Li upper limits, and open circles are giants with Li detections. The model is a YREC isochrone with [Fe/H]= +0.17 and an age of 1.66 Gyrs. Right panel: C/N ratio of the Trumpler 20 giants vs. their $T_{\rm eff}$. The solid line is a YREC stellar model of 1.8 $M_\odot$ with [Fe/H]=+0.17. We also show the Lagarde et al. (2012) models used by S16: standard models (L12s) are dotted lines in blue for a mass of 1.5 $M_\odot$ and in red for 2.0 $M_\odot$, and models with rotational and thermohaline mixing (L12rt) are dashed lines. The horizontal lines at lower $T_{\rm eff}$ in these models are due to the horizontal branch. Credit: Aguilera-Gómez et al. (2016a)

of less evolved stars in the same cluster rather than subgiants of other clusters to fix the zero-point of our RGB models. However, there is only one subgiant reported in Trumpler 20 with measured abundance (MG 430), and using only this star to calibrate our models is risky.

Considering this new interpretation of the general population of stars in Trumpler 20, the two giants with higher Li abundances would stand out as anomalous and would require additional noncanonical physics to be explained. Thus, we investigate if their high abundance could be related to engulfment of planets or brown dwarfs.

The difference in interpretation between this work and S16 is partially based on the exact evolutionary stage of the giants. The $T_{\rm eff}$-log g plane in Fig. 4.1 (left panel) shows the location of the Trumpler 20 giants, and a Yale Rotating Evolutionary Code (YREC, Pinsonneault et al. 1989; Demarque et al. 2008) model with 1.66 Gyr and [Fe/H]= +0.17. The group of giants with low log g could either be AGB or RGB, while the group of stars with high log g could be either first ascent RGB or He-core burning giants.

The C/N ratio can somehow help to solve these ambiguities in evolutionary stage obtained only with HR diagram position. During the FDU, the carbon abundance increases and the nitrogen decreases, reducing the C/N abundance ratio. We can see the effect of the FDU in the right panel of Fig. 4.1, where all of the Trumpler 20 giants have low C/N ratios, having already completed their FDU. The C/N ratio is compared with models from Lagarde et al. (2012) used by S16, in their canonical version or with rotational mixing in the MS plus thermohaline mixing in the RGB.

Additionally, our YREC standard model for a $1.8\,M_{\odot}$, [Fe/H]=+0.17 star is the black solid line.

The large typical uncertainties in C/N are the same size as the range of C/N predicted by the models, both with and without mixing, so the C/N alone does not allow us to discriminate if there is extra-mixing in Trumpler 20. We do want to mention that the error bars reported in previous works seem to be too big compared to the dispersion of the data.

The main difference between our canonical models and those of Lagarde et al. (2012) is the shift in temperature due to the use of a different mixing length parameter. We have calibrated this parameter to fit the upper RGB of the cluster, obtaining a value of $\alpha = 2.1$, slightly bigger than the solar-calibrated value of $\alpha = 1.95$. Taking into account this difference, our YREC models predict hotter temperatures at the beginning of the RGB, with all the giants in Trumpler 20 being consistent with first ascent RGB.

It is still possible that some of the stars in the cluster are burning He in their core, specifically in the group of stars with higher temperatures. We estimate a ratio between first ascent to core-He burning giants in the high-temperature group of stars, which share $\log g$ from $\log g = 3.2$ to 2.7. Based on theory, to obtain the number of RGB stars to horizontal branch stars, we compare the timescales that stars spend in each of these stages:

$$\frac{N_{\rm RGB}}{N_{\rm HB}} = \frac{t_{\rm RGB}}{t_{\rm HB}}. \tag{4.1}$$

Our YREC models for a star of $1.8\,M_{\odot}$ and [Fe/H]= +0.17 predict a time in the RGB of $t_{\rm RGB} \sim 7.5 \times 10^7$ yr, and the time in the core-He burning phase is $t_{\rm HB} \sim 1.1 \times 10^8$ yr. With this, we calculate that $\sim$ 19 of the 31 giants in the higher temperature group belong to the horizontal branch. The exact timescales can change with different assumptions and models, so this fraction should only be used as an indicative. Regardless of how this ratio changes, it shows that not all of the hotter stars in the cluster are burning He in their core. In particular, the Li-rich red giant in the high-temperature group stands out in the HR diagram, indicating that it could be an RGB star.

Li abundances of Trumpler 20 giants as calculated by the *Gaia*-ESO collaboration are compared to other similar clusters and models in Fig. 4.2. These included non-LTE corrections using the grids of Lind et al. (2009a). The left panel also includes our canonical YREC model for the $1.8\,M_{\odot}$ star but calibrated to the turnoff abundance of three open clusters. The open clusters considered are NGC 752 (Böcek Topcu et al. 2015), NGC 3680 (Pasquini et al. 2001; Delgado Mena et al. 2016), and IC 4651 (Pasquini et al. 2004; Delgado Mena et al. 2016), and their Li measurements are also included in the figure. These have similar turnoff masses to Trumpler 20 but different metallicities.

NGC 752 has a similar age, with [Fe/H]~ -0.05 (Anthony-Twarog et al. 2009). We have decided to include this cluster in our comparison given its similar properties even if there is not enough data at the turnoff that can help us to calibrate models.

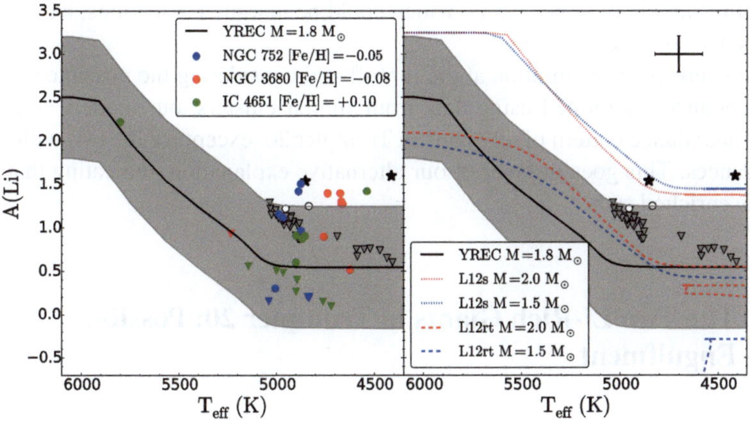

Fig. 4.2 Li abundance of Trumpler 20 giants vs. T_{eff}. Filled triangles show upper limits, open circles are detections, and filled stars are Li-rich giants. Left panel: Trumpler 20 giants compared with other 3 open clusters. Our YREC model is also included for comparison. Given the data from these clusters, we calculate a possible spread in Li abundance in the turnoff and show it as a gray band. Right panel: Trumpler 20 giants are compared with our YREC and Lagarde et al. (2012) models (same symbols as Fig. 4.1). Giants in the cluster are consistent with the expected Li abundance once the turnoff is calibrated to other similar clusters. No extra-mixing is required during the RGB to explain the abundances of upper limits. Credit: Aguilera-Gómez et al. (2016a)

The cluster IC 4651 has a higher metallicity of [Fe/H]=+0.10 and Li abundances in the turnoff region that range from A(Li) $\sim$ 1.6 to meteoritic values. Mixing taking place before this point of evolution can explain the large spread in Li (Pasquini et al. 2004). A similarly large spread can be seen in the cluster NGC3680, with a metallicity of [Fe/H]~ -0.08 (Anthony-Twarog et al. 2009). In this case, Li turnoff abundances vary from A(Li) $\sim$ 1.9 to meteoritic values (Pasquini et al. 2001; Anthony-Twarog et al. 2009). Considering these values, we calculate an average Li abundance at the turnoff of A(Li) $= 2.5 \pm 0.7$, which we use to calibrate our models. The internal spread in cluster abundances is shown in Fig. 4.2 as a gray band, which shows all the possible abundances that giants can have once the Li is calibrated, but considering canonical evolution during the entire RGB phase.

By comparison to these other clusters, the Trumpler 20 giants do not seem to be special. It is interesting that some of the giants in these other clusters also seem to have higher abundances than what is expected canonically, even after calibrating the turnoff. These unusual stars, including the two Li-rich giants from Trumpler 20, are the points located above the gray bands in Fig. 4.2 and could be very interesting to study the enrichment mechanism in giants, including both planet accretion and extra-mixing. On the other hand, the Li upper limits are comparable to the measurements in other open clusters. It is possible that those giants located after the luminosity function bump have some extra-mixing acting, as they seem to have systematically lower abundances. However, extra-mixing is not needed to explain the entire sample of upper limits, which could be marginal detections, based on the abundances of other clusters.

With zero-point calibration at the turnoff and considering the possible spread in Li abundances calibrated using data from similar clusters, our models agree with the Li abundance pattern of all giants in Trumpler 20, except for the two with higher abundances. This goes in favor of our alternative explanation, indicating that those are the enriched stars.

4.2 The Two Li-Rich Giants in Trumpler 20: Possible Engulfment

If we consider some MS Li depletion and calibrate our models with the Li abundance at the turnoff of some clusters, all giants in Trumpler 20, except for the two Li-rich, are consistent with standard FDU and canonical RGB evolution. With this interpretation of the data in mind, the Li-rich giants followed a similar evolution to the rest of the clusters and were subsequently enriched in Li by a noncanonical mechanism such as extra-mixing or planet engulfment. A thorough analysis of the Li enrichment by substellar mass companion accretion was done in Sect. 3.2 (published in Aguilera-Gómez et al. 2016b), where we model the engulfment, confirming that it can explain Li abundances up to A(Li) $= 2.2$ in giants, well above the level of the two Li-rich giants in Trumpler 20.

Here, we model the engulfment of SSC by a $1.8\,M_{\odot}$ star (turnoff mass of the cluster) with [Fe/H]=+0.17 in an attempt to explain the high Li abundance of the unusual giants. In our original modeling of the process, we found that the planetary period is not important in the final Li abundance post-engulfment. We have decided to model the accretion at a stellar age of 1.56 Gyr or $\log g \sim 3.3$. More detailed information on the assumptions of the models, other parameters included in YREC, or properties of the modeled companions can be found in Sect. 3.2.1. Main sequence extra-mixing is not explicitly included in our models. However, setting a zero-point Li abundance in the turnoff is equivalent to consider processes that can decrease the Li previous to this point in evolution.

The Li abundance in models with and without engulfment are found in Fig. 4.3. To reach the high Li abundance of the Li-rich giants, the accretion of a planet/brown dwarf of $16^{+6}_{-10}\,M_J$ is needed. The reported error is calculated based on the uncertainty of the measured Li abundance. Changing the initial Li abundance of the star to any value within the previously calculated spread found in open clusters does not produce important variations in the needed companion mass to produce the high Li signal.

We show two examples of accreted substellar companion in Fig. 4.3, a $15\,M_J$ (dashed line) and $17\,M_J$ (dotted line). Although these values appear to be particularly high for the mass of the companion, our calculated errors suggest that a smaller companion could also produce an important effect. One of the Li-rich giants is

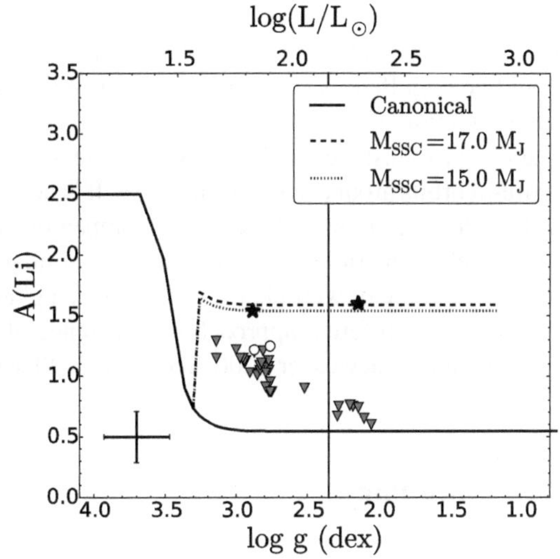

Fig. 4.3 Li abundance of Trumpler 20 giants compared to models of Li abundance for a $1.8\,M_\odot$ star that engulfs a substellar mass companion at $\log g \sim 3.3$. The solid line is the YREC model calibrated to the turnoff abundance of other similar clusters. The dashed and dotted lines are models that include the engulfment of two different companions: a $15\,M_J$ and $17\,M_J$, respectively. The vertical line marks the position of the RGB bump. Credit: Aguilera-Gómez et al. (2016a)

located after the luminosity function bump (at lower $\log g$ than the vertical line). As such, it is possible that its high Li abundance could be explained by internal Li production and enhanced extra-mixing. However, it may be difficult to explain those high abundances through this method, given that it is not known how extra-mixing works at the high metallicity of Trumpler 20, and some results show that mixing mechanisms are less efficient at higher metallicities (Martell et al. 2008).

Li enrichment by substellar companion engulfment is often associated with an increase in the rotation rate of the star (Siess and Livio 1999b). We note that the large masses of the suggested engulfed brown dwarfs by Trumpler 20 giants should spin up the giants. However, the low $v\sin i$ measured in these giants is not enough to rule out this scenario, as the giants could decrease their rotation rate as they evolve, still maintaining their high Li abundances. It is possible that the rotational signature is preserved for a long time in the RGB (Privitera et al. 2016), but how long depends on some assumptions in the modeling, such as the wind magnetic braking efficiency. A more detailed analysis of these two giants is needed to confirm or rule out the accretion scenario.

We can test this suggested picture of Trumpler 20 Li evolution by obtaining the Li abundance of subgiants and turnoff stars. Thus, we could calibrate the abundance and possible spread in Li that is really found in this open cluster, before stars go through the FDU. With that knowledge, we could also understand if giants with lower $\log g$ are showing the typical signal of an extra-mixing mechanism triggered at the RGB bump decreasing further their abundances or they had initially lower Li abundances in the turnoff.

Finally, if engulfment is producing the Li enhancement, we can use the relative number of Li-rich giants in Trumpler 20 to calculate a fraction of MS stars hosting high-mass companions in a host mass regime that has proven difficult to probe by the state-of-the-art techniques. Always considering that this estimate is based on small number statistics, and in consequence, our results are suggestive rather than conclusive, we calculate that a 5% of $1.8\,M_\odot$ host close brown dwarf/planets that could be engulfed by the star as it gets to the RGB. Some studies suggest a much larger fraction of metal-rich A-dwarfs hosting planets when compared to solar-type stars (e.g., Johnson et al. 2010). Nevertheless, there seems to be a lack of planets in close-in orbits around stars this massive. It is necessary to identify if this is produced due to the accretion of the close-in companions as the stars approaches the RGB, an unlikely scenario given current models of tidal evolution (Kunitomo et al. 2011). A larger sample of well-characterized Li-rich giants in this mass range would allow to test with a different approach the existence of the so-called brown dwarf desert (Grether and Lineweaver 2006; Troup et al. 2016) at this specific mass.

4.3 Summary

Open clusters offer the possibility to study the Li enrichment problem in samples of stars with known ages and very similar masses. We study Trumpler 20, a cluster with published atmospheric parameters and Li abundances by S16. From the 40 giants observed, 2 have higher Li abundances than the rest of the sample. We propose a scenario to explain the Li abundance pattern of the cluster where giants have experience MS depletion and then standard RGB evolution, with canonical depletion by FDU. In this context, the two giants with high Li are unusual, and they require an enrichment mechanism to be explained. A possibility is that these giants have engulfed substellar mass companions of $16^{+6}_{-10}\,M_J$. Although we cannot exclude that mixing is acting on the RGB after the luminosity function bump, given the upper limits in Li that were estimated for most of the cluster stars, the extra-mixing is not needed to explain the abundance of the giants.

There are other interesting samples of Li-rich giants that could be produced by planet accretion. One of the recent interesting samples is that of Casey et al. (2016), although a better confirmation about the exact evolutionary stage reported for those giants is needed. On the other hand, we think that the lack of close-in planets orbiting subgiants does not seem to be a strong enough argument to point to the engulfment of planets, given that this would predict a large number of Li-rich giants before the RGB bump that is not observed.

To test our explanation for the Li abundance pattern of Trumpler 20, and study the mechanism behind the enrichment, we need to measure further abundances for these and other giants in the cluster. Measuring the Li abundance of subgiants would help in the calibration of RGB stars in the cluster and provide necessary clues to understand the Li in Trumpler 20.

References

C. Aguilera-Gómez, J. Chanamé, M.H. Pinsonneault, J.K. Carlberg, ApJ **829**, 127 (2016). http://doi.org/10.3847/0004-637X/829/2/127

C. Aguilera-Gómez, J. Chanamé, M.H. Pinsonneault, J.K. Carlberg, ApJL **833**, L24 (2016). http://doi.org/10.3847/2041-8213/833/2/L24

B.J. Anthony-Twarog, C.P. Deliyannis, B.A. Twarog, K.V. Croxall, J.D. Cummings, AJ **138**, 1171 (2009). http://doi.org/10.1088/0004-6256/138/4/1171

G. Böcek Topcu, M. Afşar, M. Schaeuble, C. Sneden, MNRAS **446**, 3562 (2015). http://doi.org/10.1093/mnras/stu2380

A.R. Casey, G. Ruchti, T. Masseron, et al., MNRAS **461**, 3336 (2016). http://doi.org/10.1093/mnras/stw1512

E. Delgado Mena, M. Tsantaki, S.G. Sousa, et al., A&A **587**, A66 (2016). http://doi.org/10.1051/0004-6361/201527196

P. Demarque, D.B. Guenther, L.H. Li, A. Mazumdar, C.W. Straka, Ap&SS **316**, 31 (2008). http://doi.org/10.1007/s10509-007-9698-y

P. Donati, T. Cantat Gaudin, A. Bragaglia, et al., A&A **561**, A94 (2014). http://doi.org/10.1051/0004-6361/201322911

D. Grether, C.H. Lineweaver, ApJ **640**, 1051 (2006). http://doi.org/10.1086/500161

J.A. Johnson, K.M. Aller, A.W. Howard, J.R. Crepp, PASP **122**, 905 (2010). http://doi.org/10.1086/655775

M. Kunitomo, M. Ikoma, B. Sato, Y. Katsuta, S. Ida, ApJ **737**, 66 (2011). http://doi.org/10.1088/0004-637X/737/2/66

N. Lagarde, T. Decressin, C. Charbonnel, P. Eggenberger, S. Ekström, A. Palacios, A&A **543**, A108 (2012). http://doi.org/10.1051/0004-6361/201118331

K. Lind, M. Asplund, P.S. Barklem, A&A **503**, 541 (2009). http://doi.org/10.1051/0004-6361/200912221

S.L. Martell, G.H. Smith, M.M. Briley, AJ **136**, 2522 (2008). http://doi.org/10.1088/0004-6256/136/6/2522

L. Pasquini, S. Randich, R. Pallavicini, A&A **374**, 1017 (2001). http://doi.org/10.1051/0004-6361:20010822

L. Pasquini, S. Randich, M. Zoccali, V. Hill, C. Charbonnel, B. Nordström, A&A **424**, 951 (2004). http://doi.org/10.1051/0004-6361:20040240

M.H. Pinsonneault, S.D. Kawaler, S. Sofia, P. Demarque, ApJ **338**, 424 (1989). http://doi.org/10.1086/167210

G. Privitera, G. Meynet, P. Eggenberger, A.A. Vidotto, E. Villaver, M. Bianda, A&A **593**, A128 (2016). http://doi.org/10.1051/0004-6361/201628758

S. Randich, G. Gilmore, Gaia-ESO Consortium, The Messenger **154**, 47 (2013)

L. Siess, M. Livio, MNRAS **308**, 1133 (1999). http://doi.org/10.1046/j.1365-8711.1999.02784.x

R. Smiljanic, E. Franciosini, S. Randich, et al., A&A **591**, A62 (2016). http://doi.org/10.1051/0004-6361/201628526

G. Tautvaišienė, A. Drazdauskas, Š. Mikolaitis, et al., A&A **573**, A55 (2015). http://doi.org/10.1051/0004-6361/201424989

N.W. Troup, D.L. Nidever, N. De Lee, et al., AJ **151**, 85 (2016). http://doi.org/10.3847/0004-6256/151/3/85

References

Chapter 5
Field Giants

After several decades of studying the phenomena of Li-rich giants, there are still not many giants known with high Li abundances, and as we have discussed in previous chapters, it is not clearly understood which of these giants are actually enriched. The study of giants with high Li abundance is relevant to determine the extent of the enrichment mechanisms and identify its probable physical causes. It is clear that no matter the criteria used to select enriched objects, only a few giants in each sample have high Li abundances. The dearth of giants with high Li abundances, and our latest results, that indicate that the Li abundance of giants needs to be compared with that of other stars with similar stellar parameters to understand if they have gone through a noncanonical physical process, implies that large samples of giants are needed. We need to study the entire possible distribution of Li abundances to find which are enriched, from the very depleted giants to those with extremely high abundances.

On the other hand, studying the problem of Li enrichment in giants that are targets of planet search programs allows to study the possible engulfment scenario and other mechanisms that may need the presence of SSCs to activate. In this chapter, we report the Li abundance of 137 giants that are part of a program to find planets around evolved stars using radial velocities, and that as such, have high S/N and high-resolution spectra.

5.1 Sample Selection

The sample is comprised of 137 giants that are part of the EXoPlanets aRound Evolved StarS (EXPRESS) project (Jones et al. 2011), dedicated to finding exoplanets around giant stars. High S/N, high-resolution spectra were obtained for the target stars by this program using the Fiber-fed Extended Range Optical Spectrograph (FEROS; Kaufer et al. 1999) at the MPG/ESO 2.2m telescope at La Silla and the

C. Aguilera-Gómez, *Explaining Lithium Enriched Red Giant Branch Stars*,
Springer Theses, https://doi.org/10.1007/978-3-030-02583-0_5

echelle spectrograph on the 1.5m telescope at CTIO. The spectral resolving power provided by the FEROS spectrograph is R$\sim$ 48000, with coverage of the entire optical region, between 3500 and 9000 Å. The echelle spectrograph provides a smaller resolution of R ~45,000 covering a region between ~ 4500 and 8700 Å. The S/N ratio around the Li line (6700 Å) after combining the obtained spectra for each star ranges between 200 and 400. Details about the spectral reduction and derivation of atmospheric parameters (T_{eff}, log g, microturbulence velocity, and [Fe/H]) can be found in Jones et al. (2011). We use those obtained parameters throughout this chapter.

All of the targets are single bright stars ($V < 8$) with *Hipparcos* parallaxes (van Leeuwen 2007) with good precision, better than 14 %, with no apparent photometric variability. They are also selected by their colors, due to their position on the HR diagram, with $0.8 \leq B - V \leq 1.2$. The visual magnitudes are corrected for extinction using the 3D maps by Arenou et al. (1992) to compute absolute magnitudes.

These absolute magnitudes are used, in turn, to obtain masses and evolutionary states of the giants in the sample with isochrone fitting. The star is located on an HR diagram by using its absolute magnitude and temperature, and its position is then compared to theoretical isochrones, obtained from stellar evolutionary models. This is a powerful technique to obtain the age and properties of stellar clusters, but one that can be complicated to apply on field giants even with known distances from parallaxes, given that several isochrones of different masses in an HR diagram can be drawn near a given point. To reduce the number of isochrones that go near the point, it is also important to know the [Fe/H] of the star.

As we have seen in previous chapters, mass can be a critical parameter to understand the Li abundance of giants. We use the masses reported by Jones et al. (2011) in this work, but given the importance of this parameter, we recalculated them using a different set of stellar evolutionary tracks (Yonsei-Yale isochrones; Yi et al. 2001) and a different method to choose the closest isochrone to the point in the HR diagram. We find a very good agreement for the masses of all the giants, thus ensuring that the mass does not depend on the models used.

Most of the stars in the sample (109) are first ascending RGB stars, but some of them could be classified as horizontal branch (HB) giants. We report the evolutionary stage determination of Jones et al. (2011) for the giants, but it is relevant to notice that although those classified as RGB can be classified as such solely by their position in the HR diagram and its comparison to the isochrones, this is not the case for the most luminous RGB stars and those classified as HB. For larger luminosities, the RGB and HB stars can have a similar position in the HR diagram, and even with knowledge on the metallicity, it is difficult to determine without ambiguity which isochrone better describes the star, and it could be as easily a more massive RGB star, or a less massive HB star with the same metallicity. The evolutionary states reported in Jones et al. (2011) are finally chosen for those stars by evolutionary timescale arguments, comparing the time stars spend on the different regions of the HR diagram, and determining where it is most likely found.

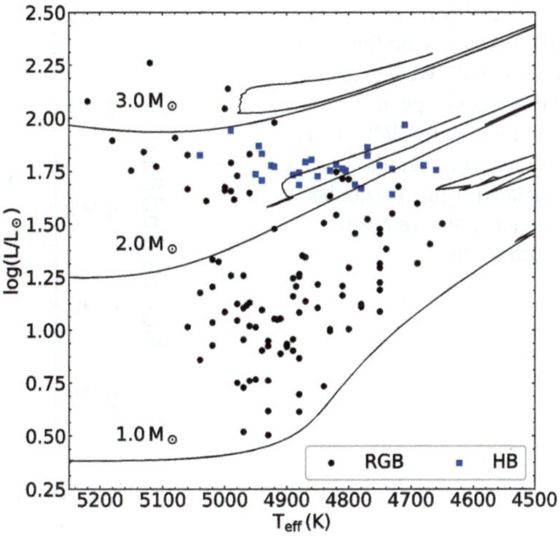

Fig. 5.1 HR diagram with the field giants in our sample. Black circles are RGB stars and blue squares are possible HB stars. Evolutionary tracks from MIST (Choi et al. 2016) for stars of masses 1.0, 2.0, and 3.0 M⊙, and solar metallicity are overplotted for guidance

An HR diagram of our sample, color-coded by the evolutionary state of the giants is in Fig. 5.1.

5.2 Measuring the Li Abundance of the Sample

We measure the Li abundance in the giants by using the line analysis and spectral synthesis code MOOG[1] (Sneden 1973, 2016 version) and the driver "synth" to produce a synthetic spectrum around the Li doublet at 6707.8 Å. In contrast with what can be done with other chemical elements, the nearby iron and CN lines in the same spectral region as the Li doublet do not allow a measurement of the Li abundance based on its equivalent width. Instead, it is necessary to produce synthetic spectra of the region around the Li doublet in 6707.8 Å to compare with the data. In Fig. 5.2 the spectral region close to the Li doublet is depicted, with the location of some spectral lines indicated. This corresponds to the fit for the giant HIP98482.

We produce synthetic spectra between 6704 Å and 6711 Å, computing model atmospheres with ATLAS9 (Kurucz 1993) to reproduce the structure of the atmosphere under the condition of local thermodynamic equilibrium, for which the $T_{\rm eff}$, $\log g$, microturbulence velocity, and [Fe/H] (from Jones et al. 2011) are needed. Besides the model atmosphere, a list of lines in the spectral region needs to be included, so that the code can identify the lines in the spectra, their position, and their intensity according to the input abundances for the respective elements. We use the molecular and atomic line lists provided by Meléndez et al. (2012) and Carlberg et al. (2012). Both of them include the wavelength, the line excitation potential, and

[1] www.as.utexas.edu/ chris/moog.html.

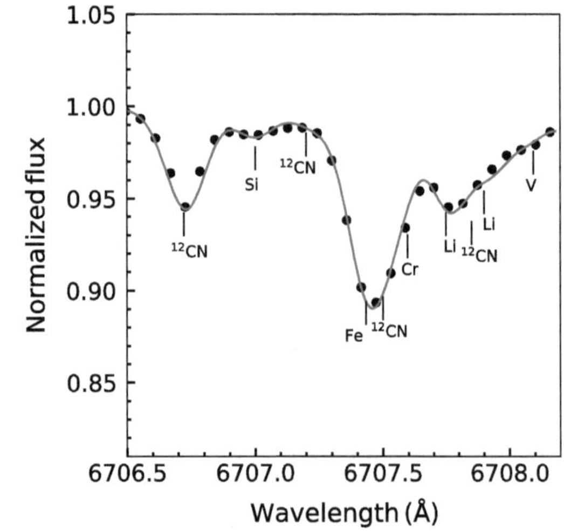

Fig. 5.2 Spectra of the giant HIP98482 around the Li line at 6707.9 Å. We can also see the synthetic spectra generated with MOOG (see text for more details) that fits the data for this particular star, and some relevant spectral lines

the $\log(gf)$ values needed, but the line list of Carlberg et al. (2012) is designed specifically to study giants, so when a line is reported in both works, we use the values from this last work.

The surface abundances of Li are obtained by minimizing the difference between the synthetic spectra and the measured spectra. To obtain the best fit, not only the Li abundance is modified but also the carbon, nitrogen, vanadium, and silicon abundances, the line broadening, and the location of the continuum. The carbon and nitrogen abundances were modified, but we kept the C/N ratio near the value found for giants after first dredge up, between C/N$\sim$ 1.0 and C/N$\sim$ 1.5. The 21 giants in the sample that have detections of the Li lines are depicted in Fig. 5.3, where we see both the fitted spectra and the data.

Two different types of error must be considered for the Li abundance detections. The first is given by what is considered a good fit, and it is related to the position of the continuum. The second is the systematic differences due to uncertainties in the atmospheric parameters, calculated according to the following procedure: To compute the errors in $\log g$, we recalculated the Li abundance by creating synthetic spectra of the region around the Li line, keeping the $T_{\rm eff}$ and metallicity constant but changing $\log g$ by its uncertainty to $\log g$+$\sigma_{\log g}$ and then to $\log g$-$\sigma_{\log g}$. Then, the error due to changes in $\log g$ is the average of the differences between the new calculated Li abundances and the original value. The same procedure is used to calculate errors due to changes in $T_{\rm eff}$ by 100 K and [Fe/H] by 0.2 dex. We find that changes in $\log g$ and [Fe/H] produce negligible variations in Li abundance but that changes in $T_{\rm eff}$ produce a considerable effect in the final abundance and can change it by over 0.2 dex. We add in quadrature both types of error, and we report this total value as the uncertainty in Li abundance in Table 5.1. From this analysis and the fitting procedure of the spectra, it is clear that the strength of the Li line depends greatly on $T_{\rm eff}$. To exemplify this, Fig. 5.4 shows two giants with the same Li abundance but different atmospheric parameters. This dependence on

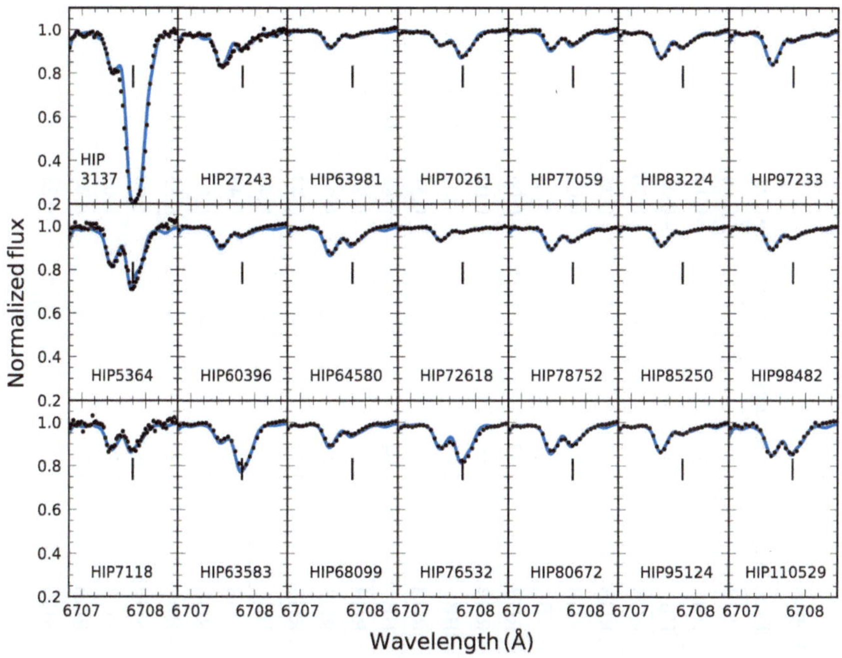

Fig. 5.3 Fit of generated synthetic spectra (blue solid line) to observed spectrum for 21 field giants. The name of the stars can be found in each panel, and the approximate position of the Li line is indicated with a short vertical line

temperature is something that also affects upper limits, as the same Li abundance in a stellar spectrum produces a stronger line if the star is cooler, and thus it is easier to detect.

In those spectra where the Li line could not be directly detected, we calculate an upper limit for the abundance. This is calculated by obtaining a limit for the equivalent width that can be measured considering a given spectrum and that is later translated into a Li abundance upper limit. This minimum equivalent width is given by the Cayrel formula (Cayrel 1988)

$$\langle \delta W \rangle^{1/2} \sim 1.5(\mathrm{FWHM}\,\delta x)^{1/2} \times \frac{1}{\mathrm{S/N}}, \tag{5.1}$$

where δx is the pixel size. To be more conservative with our upper limits, we have considered 3 times this minimum value, and calculate the abundance accordingly. We have also fitted synthetic spectra to the region, similar to the procedure followed for the detections, to ensure the upper limits were correct according to the spectrum used. Figure 5.5 shows two different giants with very similar atmospheric parameters, one with a clear detection of Li abundance and the other with only an upper limit.

Table 5.1 Stellar parameters (Jones et al. 2011) and Li abundances for the 137 sample giants

HIP	T_{eff} (K)	log g (cm s^{-2})	[Fe/H]	log $L/L_{\odot}$	Mass ($M_{\odot}$)	Evol. Stage	vsini (km/s)	Flag[a]	A(Li)	Error A(Li)
242	4990	3.14	−0.04	1.257	1.84	RGB	1.53	<	0.46	–
343	4790	2.51	0.14	1.687	1.95	HB	1.35	<	−0.11	–
655	4750	2.59	0.12	1.778	1.95	HB	1.48	<	−0.08	–
873	4920	3.06	0.07	1.478	2.08	RGB	2.72	<	0.70	–
1230	4880	2.95	−0.09	1.264	1.55	RGB	1.62	<	0.35	–
1684	4970	3.16	0.28	1.104	1.83	RGB	1.87	<	0.6	–
3137	4730	2.57	0.05	1.642	1.85	HB	2.81		1.7	0.18
3436	4750	2.9	0.17	1.189	1.47	RGB	2.09	<	−0.06	–
4618	4750	2.91	0.01	1.294	1.45	RGB	1.56	<	−0.02	–
5364	4770	2.96	0.24	1.863	2.38	HB	2.64		1.01	0.20
6116	4850	3.11	0.04	1.105	1.53	RGB	1.51	<	0.31	–
7118	4820	2.74	−0.06	1.783	1.85	HB	2.04		0.81	0.28
8541	4670	2.7	−0.15	1.405	1.15	RGB	1.28	<	−0.3	–
9406	4950	3.37	−0.04	0.767	1.25	RGB	1.4	<	0.24	–
10164	4930	3.3	0.14	0.762	1.36	RGB	1.72	<	0.43	–
10234	4940	2.59	−0.17	1.833	1.85	HB	1.27	<	0.43	–
10548	4980	3.36	0.11	1.045	1.66	RGB	1.64	<	0.48	–
11600	4970	3.62	0.33	0.519	1.3	RGB	2.14	<	0.6	–
11791	4890	2.68	0.01	1.733	2.09	HB	1.22	<	0.33	–
13147	4820	2.45	−0.37	1.747	1.53	RGB	2.19	<	0.43	–
16989	4960	2.59	0	1.831	2.61	RGB	2.19	<	0.62	–
17183	4930	3.41	−0.07	0.617	1.05	RGB	1.3	<	0.4	–
18056	4820	2.83	−0.17	1.544	1.55	RGB	1.53	<	0.28	–

18606	4950	3.19	0.07	1.014	1.61	RGB	1.52	<	0.42	–
19483	5080	2.72	0.13	1.907	2.99	RGB	3.19	<	0.68	–
21154	4780	3	0.18	1.122	1.54	RGB	1.59	<	0.08	–
21685	4650	2.55	−0.31	1.503	0.96	RGB	2.51	<	0.06	–
21743	4990	2.82	−0.36	1.942	2.08	HB	1.57	<	0.67	–
22479	4990	2.93	0.11	1.79	2.42	RGB	2.41	<	0.66	–
22491	5000	3.23	−0.16	1.085	1.45	RGB	1.53	<	0.55	–
23067	5020	3.2	−0.13	1.332	1.85	RGB	1.31	<	0.52	–
24130	4910	3.22	−0.02	0.889	1.35	RGB	1.53	<	0.25	–
24275	4890	3.05	0.17	1.159	1.74	RGB	1.65	<	0.3	–
26019	4690	2.54	0	1.598	1.63	RGB	1.83	<	−0.05	–
26649	5220	2.94	0.08	2.079	3.11	RGB	3.14	<	0.96	–
27243	5120	2.81	0.23	2.262	3.48	RGB	3.8		1.18	0.19
27434	4830	3.08	0.13	0.996	1.45	RGB	1.6	<	0.23	–
33139	5060	3.36	0.15	1.015	1.66	RGB	1.92	<	0.8	–
35154	4930	3.5	0.35	0.504	1.27	RGB	2.19	<	0.6	–
39738	4980	3.26	−0.06	1.046	1.46	RGB	1.58	<	0.39	–
41683	4980	3.35	0.2	0.751	1.43	RGB	1.85	<	0.65	–
41856	4900	3.18	0.17	0.935	1.54	RGB	1.81	<	0.22	–
56260	4890	3.12	0.23	0.903	1.53	RGB	2.24	<	0.28	–
56640	4780	2.94	0.09	1.109	1.45	RGB	1.8	<	−0.1	–
58782	4810	2.81	−0.15	1.208	1.25	RGB	1.68	<	0.04	–
59016	4800	2.88	0.07	1.294	1.64	RGB	1.87	<	−0.04	–
59367	4960	3.08	−0.01	1.02	1.57	RGB	2.04	<	0.42	–
60035	4890	3.17	0.23	0.957	1.55	RGB	2.22	<	0.33	–
60374	4940	2.6	−0.04	1.708	2.11	HB	1.44	<	0.4	–

(continued)

Table 5.1 (continued)

HIP	T_{eff} (K)	log g (cm s^{-2})	[Fe/H]	log $L/L_{\odot}$	Mass ($M_{\odot}$)	Evol. Stage	vsini (km/s)	Flag[a]	A(Li)	Error A(Li)
60396	4810	2.52	−0.18	1.762	1.55	HB	1.35		0.42	0.28
62447	4990	2.88	0.13	1.657	2.46	RGB	2.4	<	0.39	–
63242	4830	2.53	−0.31	1.633	1.54	RGB	3.65	<	0.24	–
63243	4880	2.57	−0.08	1.743	1.75	HB	1.39	<	0.27	–
63583	4800	2.46	−0.22	1.713	1.65	RGB	1.8		1.25	0.29
63981	4840	2.85	−0.22	1.506	1.55	RGB	1.8		0.38	0.19
64580	4770	2.55	0.02	1.825	2.13	HB	1.31		0.32	0.24
64590	4870	2.7	0.18	1.794	2.4	HB	1.74	<	0.13	–
64647	4870	2.92	0.01	1.343	1.82	RGB	2.46	<	0.2	–
64803	5060	2.63	0.04	1.827	2.41	RGB	2.4	<	0.59	–
65238	4810	2.82	0.14	1.163	1.56	RGB	2.11	<	−0.05	–
65373	4920	2.75	0.15	1.979	2.94	RGB	2.1	<	0.3	–
65891	5000	2.9	0.16	1.66	2.46	RGB	2.52	<	0.48	–
66427	5180	3	0.12	1.894	2.97	RGB	2.33	<	0.73	–
66711	5000	3.22	0.13	1.086	1.75	RGB	2.2	<	0.45	–
66924	4860	2.53	−0.18	1.805	1.55	HB	1.39	<	0.26	–
67537	4985	2.85	0.15	1.617	2.43	RGB	2.34	<	0.48	–
67851	4890	3.15	0	1.244	1.67	RGB	1.84	<	0.29	–
67890	4750	2.81	0.31	1.313	1.73	RGB	2.33	<	−0.08	–
68054	5110	2.9	0.13	1.772	2.71	RGB	3.46	<	0.88	–
68099	5130	3	0.15	1.841	2.93	RGB	2.8		1.1	0.17
68263	4870	3.05	−0.11	1.136	1.35	RGB	1.64	<	0.27	–
68333	4925	2.55	−0.32	1.777	1.55	HB	1.39	<	0.44	–

69065	4960	2.62	−0.22	1.648	1.95	RGB	2.32	<	0.39	–
70261	4810	2.65	−0.38	1.715	1.52	RGB	1.33		1.05	0.18
70514	4750	2.65	−0.12	1.454	1.35	RGB	1.71	<	−0.17	–
70987	4880	2.64	−0.03	1.686	1.95	HB	1.18	<	0.28	–
71778	5040	3.45	0.03	0.86	1.51	RGB	1.83	<	0.58	–
72097	5000	2.72	−0.03	1.675	2.35	RGB	2.42	<	0.58	–
72618	4930	3.14	−0.28	0.925	1.14	RGB	1.32		0.47	0.13
73758	4840	3.2	0.41	0.735	1.4	RGB	2.83	<	0.36	–
74188	4750	2.95	0.12	1.087	1.36	RGB	1.8	<	−0.13	–
74239	5000	2.77	0.07	2.045	3.07	RGB	3.93	<	0.92	–
74890	4850	3.06	0.2	1.215	1.74	RGB	2.23	<	0.15	–
75092	4940	3.17	0.09	1.095	1.66	RGB	2.02	<	0.35	–
75101	4880	3.35	0.29	0.614	1.27	RGB	2.09	<	0.4	–
75331	4880	3.33	0.31	0.696	1.38	RGB	2.17	<	0.26	–
76532	4850	2.77	0.02	1.727	1.97	HB	1.44		1.02	0.29
76569	4830	2.78	−0.18	1.754	1.55	HB	2.42	<	0.21	–
77059	5010	3.14	0.01	1.322	1.98	RGB	1.56		1.01	0.18
77888	4690	2.63	0.02	1.314	1.35	RGB	1.93	<	−0.23	–
78752	4970	3.47	0.09	0.728	1.38	RGB	1.56		0.94	0.18
78868	4660	2.38	0.3	1.756	1.95	HB	1.95	<	−0.2	–
80672	4710	2.47	0.03	1.968	2.29	HB	1.39		0.25	0.19
80687	5020	3.3	0.03	0.928	1.54	RGB	1.39	<	0.59	–
82653	4790	2.8	0.04	1.456	1.75	RGB	1.73	<	−0.06	–

(continued)

Table 5.1 (continued)

HIP	T_{eff} (K)	log *g* (cm s^{-2})	[Fe/H]	log $L/L_{\odot}$	Mass ($M_{\odot}$)	Evol. Stage	vsini (km/s)	Flag[a]	A(Li)	Error A(Li)
83224	4880	2.91	0.07	1.253	1.75	RGB	1.71		0.72	0.27
84056	4960	3.17	0.08	1.129	1.75	RGB	1.67	<	0.39	–
84248	4730	2.42	−0.12	1.761	1.25	HB	2.98	<	−0.27	–
85250	4980	3.15	−0.17	1.124	1.45	RGB	1.4		0.55	0.18
86208	4730	2.73	−0.16	1.551	1.35	RGB	1.44	<	−0.11	–
86248	4680	2.28	0.02	1.777	1.55	HB	1.74	<	−0.36	–
86368	4880	3.13	0.12	0.867	1.43	RGB	1.7	<	0.27	–
86786	4970	3.07	−0.12	1.257	1.65	RGB	1.52	<	0.4	–
87273	4750	2.64	0.21	1.224	1.56	RGB	2.43	<	−0.19	–
88684	4940	3.24	0.04	0.904	1.45	RGB	1.61	<	0.35	–
90124	4950	2.73	0.09	1.737	2.4	HB	1.82	<	0.31	–
90988	4910	3.21	0.24	1.054	1.66	RGB	1.9	<	0.34	–
92367	5040	2.68	−0.43	1.826	1.87	HB	3.11	<	0.77	–
95124	5040	3.28	0.2	1.176	1.87	RGB	1.9		0.85	0.18
95532	4970	3.2	−0.04	0.955	1.44	RGB	1.93	<	0.43	–
97233	5020	3.26	0.29	1.204	1.96	RGB	2.19		0.79	0.19
98482	4720	2.51	−0.17	1.678	1.54	RGB	1.42		0.16	0.28
98575	5150	2.81	0.15	1.753	2.44	RGB	2.92	<	0.67	–
99171	4830	3.07	−0.01	1.004	1.35	RGB	1.51	<	0.15	–
100062	4920	2.57	−0.09	1.772	1.85	HB	1.52	<	0.31	–
101477	4980	2.78	0.05	1.729	2.46	RGB	2.02	<	0.4	–
101911	4885	2.97	0.03	1.206	1.74	RGB	1.79	<	0.24	–

103836	4740	2.89	−0.06	1.382	1.45	RGB	3.23	<	0.05	–
104148	4805	2.45	0.03	1.754	1.96	HB	2.12	<	−0.22	–
104838	4900	3.18	0.04	0.928	1.45	RGB	1.44	<	0.32	–
105854	4780	2.94	0.31	1.67	2.12	HB	3.71	<	0.11	–
105856	4915	3.08	0.09	1.05	1.55	RGB	1.86	<	0.29	–
106055	4770	2.68	0.15	1.524	1.95	RGB	2.31	<	−0.05	–
106922	4875	2.96	0.12	1.351	1.86	RGB	1.62	<	0.27	–
107122	4965	3.27	0.1	1.116	1.75	RGB	1.72	<	0.48	–
107773	4945	2.59	0.03	1.869	2.46	HB	1.95	<	0.36	–
108543	4995	2.41	0.05	2.138	3.35	RGB	2.53	<	0.7	–
109228	4960	3.31	0.02	0.761	1.37	RGB	1.65	<	0.57	–
110391	4750	2.69	−0.18	1.477	1.35	RGB	1.43	<	0.17	–
110529	5060	2.8	0.13	1.666	2.46	RGB	2.6		1.38	0.20
111515	5030	2.97	0.14	1.61	2.45	RGB	2.05	<	0.75	–
111909	4930	3.24	0.23	0.949	1.57	RGB	2.25	<	0.38	–
113779	5020	3.42	0.11	1.036	1.68	RGB	1.55	<	0.5	–
114408	4880	3.12	−0.25	1.083	1.15	RGB	1.92	<	0.5	–
114933	4920	3.09	0.14	1.053	1.65	RGB	1.75	<	0.34	–
116630	4900	3.18	0.16	0.922	1.47	RGB	2.59	<	0.31	–
117411	4800	3.04	0.17	1.005	1.45	RGB	1.87	<	0.04	–

[a]Limit flag on A(Li), with < indicating upper limits

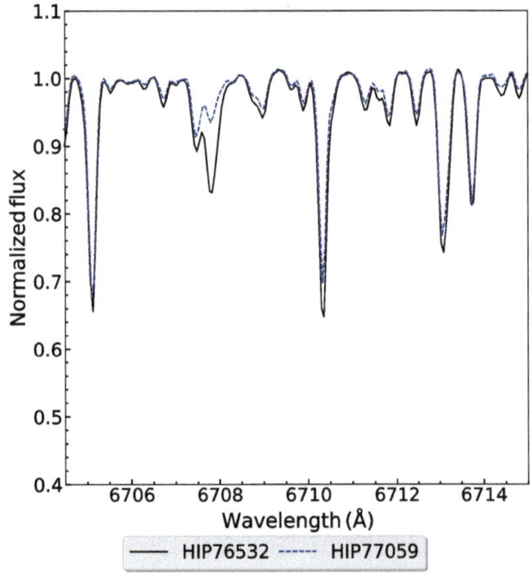

Fig. 5.4 Example of spectra for two giants with the same Li abundance but different atmospheric parameters, to show how the $T_{\rm eff}$ affects the measuremen of Li. HIP77059 is only 160 K hotter than HIP76532, both with A(Li) $\sim$ 1.01, but we can see that the line is stronger in the cooler star

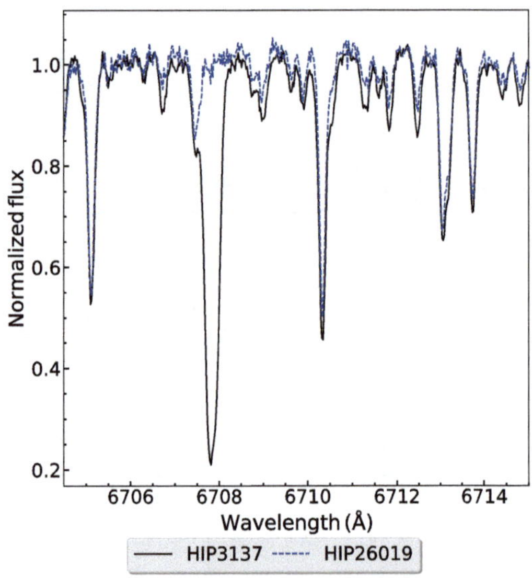

Fig. 5.5 Example of spectra for giants with similar atmospheric parameters but different Li abundances. HIP3137 is a clear detection, with a measured abundance of A(Li) $= 1.7 \pm 0.18$, while for HIP26019 we can only obtain an upper limit of A(Li) < -0.05

Calculated Li abundances, their respective errors, and upper limits, along with parameters and evolutionary stages from Jones et al. (2011), can be found in Table 5.1.

5.3 Comparison with Other Works

We have found some stars in common between our sample and large catalogs of Li abundance measurements in the literature and have compared the obtained abundance. We have one star in common with the catalog of 145 stars of Lèbre et al. (2006). This is HIP27243, for which we obtain A(Li) $= 1.18 \pm 0.19$, in contrast with the value A(Li) $=$ 0.80 of that work. We obtain a higher abundance, even considering the uncertainty in the measurement. This is due to the big difference between the effective temperature used in this work (5120 K) and the value used by Lèbre et al. (2006) of 4900 K. Not related with the value of Li abundance, we also notice a large difference in the metallicity stated for the giant. Another reported temperature for the same star (4917 K, McDonald et al. 2012) indicates that it is possible that the T_{eff} obtained by Jones et al. (2011), and that is used in this work, is overestimated, and further analysis of the atmospheric parameters are needed to precisely calculate the Li abundance of this star.

The work of Liu et al. (2014) includes atmospheric parameters and Li abundances for 378 G-K giants. We find ten stars in common with this work, most of which are upper limits. As the upper limits reported depend on the quality of the spectra, it is not possible to compare the different values directly. The only giant that shows a detection of Li in both this work and our catalog is HIP110529, for which we report an abundance of A(Li) $= 1.38 \pm 0.20$, consistent with the value in that work (A(Li) $= 1.40 \pm 0.14$). For three of the giants in common, we obtain an upper limit, while they report a very clear detection of the line. For HIP873 we have a very noisy spectrum that does not allow a very stringent upper limit (A(Li) < 0.7), much higher than the detection obtained by Liu et al. (2014). A similar case is that of the giant HIP111515, where we get an upper limit (A(Li) < 0.75), and they report a detection of A(Li) $= 0.51$. The case of the star HIP16989 is different. We report an upper limit of 0.62, while this is a clear detection of the same value for Liu et al. (2014). Given the difference in atmospheric parameters and spectral quality, we think that the differences between this work and the catalog of Liu et al. (2014) are within acceptable limits.

When comparing with the catalog of Luck and Heiter (2007), we find eight stars in common, none of which have detected Li in our sample. All of those giants are also included in the most recent catalog by Luck (2015), where we find 14 of the giants in our sample. Stars HIP5364 and HIP76532 have detections of Li in both works and agree well considering uncertainties (HIP5364: A(Li) $= 1.01 \pm 0.20$ this work, A(Li) $= 1.17$ Luck 2015; HIP76532: A(Li) $= 1.02 \pm 0.29$ this work, A(Li) $= 1.14$ Luck 2015), and for two other giants, we present upper limits that are consistent with the detections presented in Luck (2015).

5.4 Results

The distribution of Li abundances for stars in our sample can be seen in Fig. 5.6. The gray histogram shows the entire sample, while the darker gray portion of the bars represents the 21 giants in total with detections of the Li line. As we can see, every star with A(Li) > 0.7 has a Li line strong enough to be detected, while on the other end of the distribution, for the same Li abundance some are detections, and

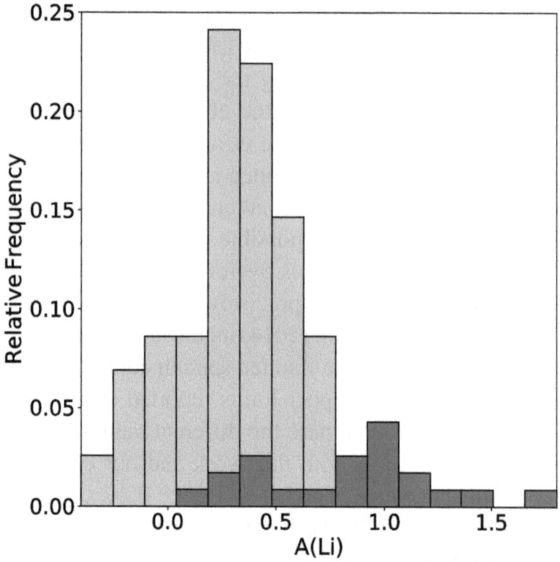

Fig. 5.6 Distribution of Li abundances for field giants in our sample. The gray histogram is the entire sample, while dark gray represents the giants with detection of the Li line

some are upper limits, depending mainly on the temperature of the giant considered, as we briefly discussed in previous sections. This is only a way to show the different abundances measured in our sample, but most of the distribution, including the peak near A(Li) $\sim$ 0.4, is formed almost entirely by upper limits, and as such it is likely to change toward lower abundances with better-quality data.

The largest Li abundance for the giants in our sample is found in the star HIP3137 that has an abundance of A(Li) $= 1.7 \pm 0.18$. This is a possible horizontal branch star with a mass of 1.85 $M_\odot$, close to the average mass of the sample.

We analyze the lithium abundance versus stellar parameters of our sample in Fig. 5.7. In the top panel, we see how the upper limits correlate with the $T_{\rm eff}$, setting the lower envelope of the distribution. As already discussed, this is an artificial effect, produced by the detection limit of the Li line for spectra with a given quality. Given a S/N, a higher Li abundance is needed in higher $T_{\rm eff}$ stars to produce the same strength of the line than for a lower temperature star. In the next panel of the figure, we see that most of the stars in the sample are metal-rich, and we see no correlation with Li abundance. Although the size of the convective envelope depends on the metallicity of the star, it also depends on its mass, so it is expected that we cannot identify correlations when considering stars of different parameters.

The third panel of Fig. 5.7 is the Li abundance as a function of the mass of the stars. There is a slight positive correlation between both quantities, with some stars that appear to have higher Li abundance than other stars of the same mass. In particular, HIP3137 with A(Li) $=$ 1.7 seems to be enriched in all panels. However, in the third panel of Fig. 5.7, several other giants (e.g., three giants with masses around 1.5 $M_\odot$) seem to be enriched when compared to stars of other similar properties that would be neglected in previous studies as they do not have

Fig. 5.7 From top to bottom, lithium abundances versus T_{eff}, metallicity, mass, and vsini. Circles are detection, while triangles are upper limits, which are also drawn in lighter colors. Stars are color-coded according to their evolutionary status; in blue are first ascending RGB stars and in green, the possible horizontal branch stars

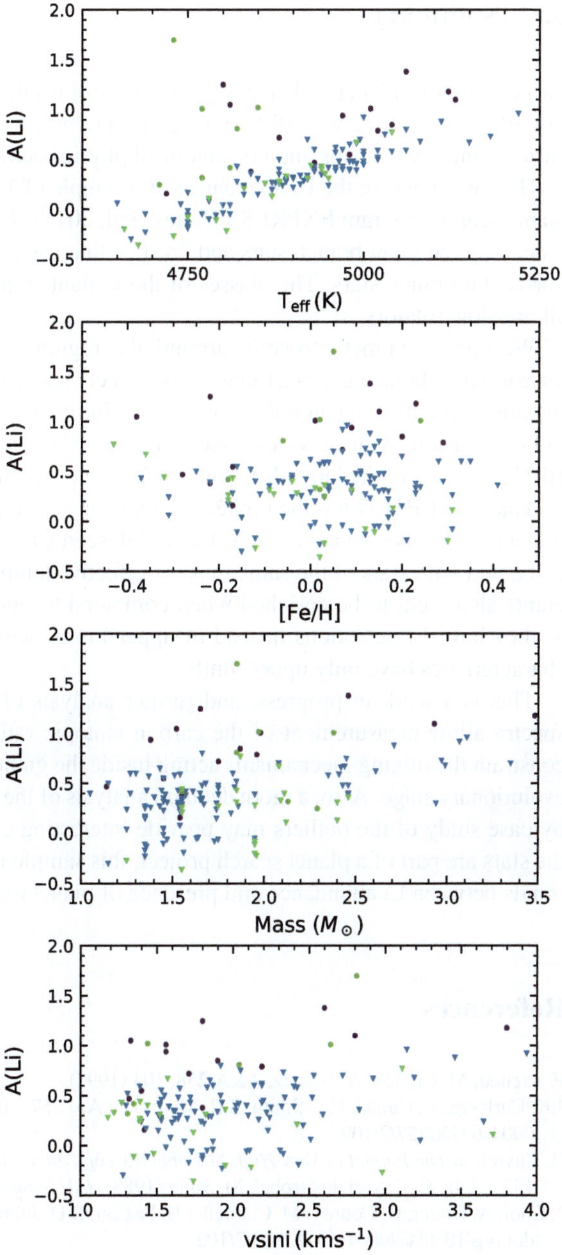

A(Li) > 1.5. The bottom panel shows A(Li) and vsini. As it could be expected, since our sample stars are selected as part of a planet search program, they all have slow rotation, so this sample is not suited to study possible correlations between fast rotation and lithium enhancement.

5.5 Summary

Large samples of Li abundance in giants are still needed to understand the distribution of Li abundance with different stellar parameters and eventually constrain the enrichment mechanisms and noncanonical physics acting on these stars.

Here we measure the Li abundance of a sample of 137 giants that are part of the planet search program EXPRESS (Jones et al. 2011). This sample includes 109 first ascending red giant branch stars and 28 stars that can either be classified as RGB or horizontal branch stars. The masses of these giants range from 1.0 to 3.5 $M_\odot$, and all are slow rotators.

We create synthetic spectra around the region of the Li line ($\sim$6708 Å) to measure the abundance. In 21 giants, we detect clearly the Li line, while for the 116 remaining giants, we can only report upper limits. Our obtained abundances seem to be consistent with previous measurements in the literature, except for the giant HIP27243, where the large difference in $T_{\rm eff}$ can explain the difference in A(Li).

The star HIP3137 has the highest Li abundance of the sample, with A(Li) = 1.7. It is a possible horizontal branch star, which seems to have a high abundance when compared with stars of the same mass or effective temperature in the sample. Other giants also seem to be enriched when compared to other giants of similar masses, as they have Li detections instead of upper limits, while other giants with similar characteristics have only upper limits.

This is a work in progress, and further analysis of the sample is needed. The spectra allow measurement of the carbon isotopic ratio that can be interesting to constrain the mixing mechanisms acting inside the giants and possibly identify their evolutionary stage. Also, a more detailed analysis of the presented trends and a case-by-case study of the outliers may provide interesting conclusions. Furthermore, as the stars are part of a planet search project, this sample is ideal for studying possible trends between Li abundance and presence of planetary companions.

References

F. Arenou, M. Grenon, A. Gomez, A&A **258**, 104 (1992)

J.K. Carlberg, K. Cunha, V.V. Smith, S.R. Majewski, ApJ **757**, 109 (2012). http://doi.org/10.1088/0004-637X/757/2/109

R. Cayrel, in *The Impact of Very High S/N Spectroscopy on Stellar Physics*, *IAU Symposium*, vol. 132, ed. by G. Cayrel de Strobel, M. Spite (1988), *IAU Symposium*, vol. 132, p. 345

J. Choi, A. Dotter, C. Conroy, M. Cantiello, B. Paxton, B.D. Johnson, ApJ **823**, 102 (2016). http://doi.org/10.3847/0004-637X/823/2/102

M.I. Jones, J.S. Jenkins, P. Rojo, C.H.F. Melo, A&A **536**, A71 (2011). http://doi.org/10.1051/0004-6361/201117887

A. Kaufer, O. Stahl, S. Tubbesing, P. Nørregaard, G. Avila, P. Francois, L. Pasquini, A. Pizzella, The Messenger **95**, 8 (1999)

R. Kurucz, ATLAS9 Stellar Atmosphere Programs and 2 km/s grid. Kurucz CD-ROM No. 13. Cambridge, Mass.: Smithsonian Astrophysical Observatory, 1993. **13** (1993)

A. Lèbre, P. de Laverny, J.D. Do Nascimento, Jr., J.R. de Medeiros, A&A **450**, 1173 (2006). http://doi.org/10.1051/0004-6361:20053485

Y.J. Liu, K.F. Tan, L. Wang, G. Zhao, B. Sato, Y. Takeda, H.N. Li, ApJ **785**, 94 (2014). http://doi.org/10.1088/0004-637X/785/2/94

R.E. Luck, AJ **150**, 88 (2015). http://doi.org/10.1088/0004-6256/150/3/88

R.E. Luck, U. Heiter, AJ **133**, 2464 (2007). http://doi.org/10.1086/513194

I. McDonald, A.A. Zijlstra, M.L. Boyer, MNRAS **427**, 343 (2012). http://doi.org/10.1111/j.1365-2966.2012.21873.x

J. Meléndez, M. Bergemann, J.G. Cohen, et al., A&A **543**, A29 (2012). http://doi.org/10.1051/0004-6361/201117222

C. Sneden, ApJ **184**, 839 (1973). http://doi.org/10.1086/152374

F. van Leeuwen, A&A **474**, 653 (2007). http://doi.org/10.1051/0004-6361:20078357

S. Yi, P. Demarque, Y.C. Kim, Y.W. Lee, C.H. Ree, T. Lejeune, S. Barnes, ApJS **136**, 417 (2001). http://doi.org/10.1086/321795

Chapter 6
Stars with Low Main Sequence Li Abundances

An interesting sample to study the Li enrichment mechanism is giants that have evolved from the Li dip or that have evolved from any other population with very low Li abundances already in the MS or subgiant phases.

The Li desert, the region in the T_{eff}-A(Li) plane where stars separate into a low and a high Li abundance, presents a possibly interesting sample with low Li abundances previous to the RGB phase to study the Li enrichment. By analyzing these stars, we can identify the progenitor of RGB stars with extremely low Li abundances, where any noncanonical process can be more easily identified.

Not only interesting to the topic of Li-rich giants, several astrophysical phenomena can be studied by using the Li abundance of stars. But, to study all these problems and characterize stars with low Li abundance, i.e., interesting progenitors to RGB stars where substellar companion engulfment and extra-mixing can be more easily identified, a homogeneous and large sample of stars with measured Li abundance is necessary. Some works have already presented internally homogeneous catalogs of Li in main sequence stars, such as Guiglion et al. (2016), who within the AMBRE project presented over $\sim$ 7200 measurements of Li. Being the largest catalog to date, its only disadvantage is that there is no way to reliably measure masses and ages for all the stars, characteristic that are relevant to understand all processes related to stellar evolution. Ideally, we would need a sample where masses can be measured. As an example, the catalog compiled by Ramírez et al. (2012) has measured parallaxes for all stars that allow to derive the masses and ages. In that work, Ramírez et al. (2012) not only present new abundances for stars but also compile literature data and homogenize it, so that all atmospheric parameters in the sample are set to the same scale. Given that several works have measured Li abundances for a large amount of stars since then, some of which do have parallaxes to measured their masses and ages. Thus, in this work we homogenize those new samples to the same scale. Of course, the future of these studies is the *Gaia* Mission, measuring parallaxes especially when combined

C. Aguilera-Gómez, *Explaining Lithium Enriched Red Giant Branch Stars*,
Springer Theses, https://doi.org/10.1007/978-3-030-02583-0_6

with spectroscopic missions that can add to the astrometric solution of stars some chemical abundances.

In this chapter, we attempt to extend the work of Ramírez et al. (2012) (R12 hereafter), that previously studied the region of the Li desert, presenting new observations of dwarfs and homogenizing pre-existing data to compile a large catalog. We follow a similar procedure, where we present new spectroscopic data for 227 stars in the temperature range of the Li desert, including their atmospheric parameters and Li abundance determinations, and we homogenize literature data to create an updated catalog. The new measurements are presented in Sect. 6.1. Literature data is homogenized following the procedure in Sect. 6.2. We present the main features of the extended catalog in Sect. 6.3. We analyze the nature of the lithium desert in Sect. 6.4 and compare our main conclusions and results with previous works and suggestions on the topic of the Li desert in Sect. 6.5. Section 6.6 contains our summary.

6.1 Target Selection and Observations

To study the Li desert, we plan observations of main sequence and subgiant stars in the $T_{\rm eff}$ range of the desert without previous measurements of Li abundance. We used the stellar parameter compilation provided by J. Meléndez (private communications) to select stars with $[{\rm Fe/H}] > -0.4$, $V < 8$, and in the $T_{\rm eff}$ range from 5800 to 6350 K. We allowed a larger range of effective temperatures to consider possible systematics in parameters found by previous works in the region of the desert. Additionally, we select only targets that were not part of the R12 catalog. Given the possible connection of the stars in the Li desert with their orbiting planets, we also program the observation of known planet hosts.

We obtain high-resolution spectroscopic observations from both the MIKE spectrograph on the 6.5-m Clay Magellan Telescope at Las Campanas Observatory (LCO) and the Tull spectrograph on the 2.7-m Harlan J. Smith Telescope at the McDonald Observatory (McD). The five-night run at LCO was carried out during July 2012, in which we observe 89 stars. Details on the reduction procedure for these spectra can be found in Section 2.2 of Ramírez et al. (2014). The two four-night runs at McD were carried out in October 2012 and March 2013, with which we obtained data for 139 stars. Details on setup and reduction for these runs are found in Section 3.1 of Chanamé and Ramírez (2012). Additionally, in all runs, we obtained a solar spectrum via the asteroid Vesta. The spectral resolution $\lambda/\Delta\lambda$ of the spectra is 60000, and the signal-to-noise ratio is typically around 400.

Stellar parameters and Li abundances are measured for the 227 stars in the sample as were found in R12. Temperatures are estimated from colors and color-temperature relations from Casagrande et al. (2010), where the uncertainty depends mainly on the color to color scatter. To measure the $\log g$ values, we use the *Hipparcos* parallaxes and isochrones, where the errors now depend mainly on the errors of parallax. This is detailed in Sect. 6.2. Metallicities are obtained by measuring the iron abundance for each line using MOOG (2014 version Sneden 1973) and comparing that value with the what is found in the solar spectra for that same line. Then, we average these differences between each line to find our final [Fe/H]. This procedure reduces systematic uncertainties related to each line.

Li abundances are also found using MOOG, now fitting a synthetic spectrum to the region around the 6708 Å Li doublet in each star. To do this, we use MARCS model atmospheres (Gustafsson et al. 2008), interpolated for the parameters of each star.

Additionally, with the relation of Fischer and Valenti (2005), we calculate the macroturbulent velocity and estimate $vsini$ when fitting the spectrum.

The derived uncertainties in parameters and Li abundances are very similar to those of R12, corresponding to $\Delta T_{\rm eff} \simeq 50\,{\rm K}$, $\Delta \log g \simeq 0.06\,{\rm dex}$, $\Delta[{\rm Fe/H}] \simeq 0.05\,{\rm dex}$, and $\Delta A({\rm Li}) \simeq 0.05\,{\rm dex}$. All of the stellar parameters for the 227 stars and uncertainties can be found in Aguilera-Gómez et al. (2018) or in electronic form at the CDS via anonymous ftp to cdsarc.u-strasbg.fr (130.79.128.5) or via http://cdsweb.u-strasbg.fr/cgi-bin/qcat?J/A+A/.

6.2 Assembling the Catalog

In this section we present the final catalog of stellar parameters and Li abundances, which includes the 227 stars with new measurements, the literature data from Delgado Mena et al. (2014, 2015), and Gonzalez (2014, 2015). Our catalog also includes the already homogeneous catalog of R12, with measurements from Lambert and Reddy (2004), Luck and Heiter (2006), Israelian et al. (2009), Baumann et al. (2010),Ghezzi et al. (2010), Gonzalez et al. (2010), and Takeda et al. (2010). At the same time, Lambert and Reddy (2004) presented data from Balachandran (1990), Chen et al. (2001), and Reddy et al. (2003).

Each of these works calculates the stellar parameters and Li in a specific way, being internally consistent. However, to generate a homogeneous catalog, we need to consider the systematic differences between works normalizing all data to our measurements. This is a similar procedure to that used by R12. We find parameters and Li with the exact same procedure used by R12, so the data for our 227 stars is consistent with the 671 stars measured in R12. Thus, to homogenize the catalog, we use as a baseline to normalize literature data a more numerous catalog of 898 stars.

To homogenize the data, we use stars in common between each sample and the combined R12+227 data and compute offsets in each parameter. These are then subtracted to the literature data, homogenizing the catalog. Although R12 present an already homogeneous catalog of stars, we re-homogenize measurement from those works again now using the combined larger sample of 898 stars, which can change the previous offsets. The offsets calculated are those in Table 6.1, originally in Aguilera-Gómez et al. (2018).

As the homogenization process is the same used in R12, details can be found there. Some relevant points:

Table 6.1 Offsets in atmospheric parameters and Li abundance between the literature parameters and our 227 measurements (TW) plus R12. Credit: Aguilera-Gómez et al., 2018, A&A, 614, A55, reproduced with permission ©ESO

Sample[a]	$\Delta T_{\rm eff}$ (K)	$\Delta \log g$	Δ[Fe/H]	ΔA(Li)	N_c[b]	N_u[c]	N_o[d]
TW+R12	0 ± 00	0.00 ± 0.00	0.00 ± 0.00	0.00 ± 0.00	-	898	835
B10	40 ± 48	0.05 ± 0.06	0.03 ± 0.03	0.07 ± 0.09	69	117	117
G10	-3 ± 69	-0.02 ± 0.04	0.04 ± 0.04	-0.14 ± 0.12	45	138	152
I09	26 ± 32	0.05 ± 0.05	0.02 ± 0.02	0.06 ± 0.04	11	79	80
LH06	46 ± 91	0.04 ± 0.20	0.01 ± 0.06	0.01 ± 0.11	77	165	194
T10	-12 ± 54	-0.03 ± 0.09	0.03 ± 0.04	-0.03 ± 0.10	26	113	117
Gh10	30 ± 71	-0.06 ± 0.14	-0.00 ± 0.06	0.01 ± 0.08	38	253	262
LR04	-83 ± 63	-0.01 ± 0.11	-0.02 ± 0.05	-0.07 ± 0.08	212	415	451
DM14	7 ± 80	0.07 ± 0.09	0.01 ± 0.04	0.04 ± 0.12	35	307	326
DM15	46 ± 70	0.12 ± 0.12	0.03 ± 0.06	0.04 ± 0.10	65	778	836
G1415	42 ± 82	0.01 ± 0.04	0.02 ± 0.05	-0.17 ± 0.07	29	66	68

[a]B10=Baumann et al. (2010); G10=Gonzalez et al. (2010); I09=Israelian et al. (2009); LH06=Luck and Heiter (2006); T10=Takeda et al. (2010); Gh10=Ghezzi et al. (2010); LR04=Lambert and Reddy (2004); DM14=Delgado Mena et al. (2014); DM15=Delgado Mena et al. (2015); G1415=Gonzalez (2014)+Gonzalez (2015)
[b]Number of stars in common between the sample and the combined data from this work and R12
[c]Number of stars of the work used in our final catalog
[d]Number of dwarfs and subgiants in the original sample

- LTE Li abundances were used.
- When measurements for a star are available in more than one work, we obtain an average of the parameters, weighting each measurement by the variances.
- Errors reported by Israelian et al. (2009) and Delgado Mena et al. (2014, 2015), although internally consistent, are much smaller than in other samples. To account for this effect, we increase their errors, as detailed in R12.
- The lower Li upper limit reported for a source is the one adopted in the catalog. In contrast, if there is a detection among upper limits, the detection is reported in the catalog.

The homogeneous catalog so far has 2484 stars with measured $T_{\rm eff}$, $\log g$, [Fe/H] and A(Li); however, to study Li we also need to remove spectroscopic binaries that can interact with their companions, modifying their Li abundances, and that can have different photometry. The Ninth Catalog of Spectroscopic Binary Orbits (Pourbaix et al. 2004) indicates that 96 of the stars in the catalog are spectroscopic binaries. It is still possible that our sample is somewhat contaminated by other types of binaries.

From the 2388 stars in the catalog, we select only those that have parallaxes so that we can measure masses and ages using theoretical isochrones. We use revised *Hipparcos* parallaxes (van Leeuwen 2007). The first Data Release of the *Gaia* mission (Gaia Collaboration et al. 2016) with the *Tycho-Gaia* Astrometric Solution (TGAS) became available while developing this work. These improved parallaxes could change measurements of age and mass, both fundamental for understanding the behavior of Li in stellar interiors. 1857 had both *Hipparcos* and TGAS parallaxes, which are generally consistent with the exception of 18 stars. These stars are peculiar and most probably binaries and are thus removed from the catalog.

Also discarding all the stars without parallaxes, for which we cannot measure consistently the mass and age, we create the final catalog with 2318 dwarfs and subgiants.

Masses, ages, and $\log g$ are calculated by placing stars on a color-magnitude diagram using their $T_{\rm eff}$ and absolute magnitudes. The position of each star in this diagram is then compared to theoretical Yonsei-Yale isochrones with metallicities fixed to the calculated metallicity of the star. Then, to choose the closest isochrone, we create probability distribution functions for the different parameters we want to find around a 3-σ radius from the observed parameters. With this method, we also take into account for the calculation of uncertainties how the different isochrones cluster in the color-magnitude diagram. The most probable parameters are adopted, along with the 1σ Gaussian-like errors, directly obtained from the probability distribution functions. Asymmetric distributions will result in asymmetric errors, as is the case for the age. More details about this process can be found in Chanamé and Ramírez (2012).

With the isochrone fitting procedure, we obtain a $\log g$ for every star in the same way. Given that this $\log g$ is internally consistent in our catalog, we prefer to use it instead of the $\log g$ found spectroscopically, calculated by each work separately.

Regarding the ages, we prefer gyrochronology ages (Barnes 2003, 2007) for the younger stars, up to 2.5 Gyr, much better constrained than ages from isochrone fitting. Gyrochronology indicates that stars with convective envelopes have a direct relationship between their age and rotation rates, which means that by measuring the period of these stars and using calibrated gyrochronology relations (Mamajek and Hillenbrand 2008), we obtain an age. Following R12, we use the rotation periods of Gaidos et al. (2000), Strassmeier et al. (2000), and Pizzolato et al. (2003). The uncertainties for the ages calculated with this method are typical of 0.1 Gyr. Additionally, throughout this work, we define stars with reliable ages as those with age/(error age)> 3.0, which constitute $\sim 47\%$ of our sample.

We show in Fig. 6.1 our final catalog. Top panel shows a color-coded by metalicity HR diagram of the sample, with $\log g$ calculated using isochrones and $T_{\rm eff}$. We see that our sample has both main sequence and subgiant stars. The middle panel shows the same HR diagram, but now stars with the $T_{\rm eff}$ of the Li desert are color-coded according to their Li abundances. Stars above the desert have high Li, while stars below the desert are Li-depleted. In the bottom panel, we see the temperature distribution of the catalog and previous literature determinations (Delgado Mena et al. 2014, 2015; Gonzalez 2014, 2015), where it is clear that our new measurements increase considerable the sample size at that $T_{\rm eff}$. It is also interesting that Delgado Mena et al. (2015) present a sample of Li abundance upper limits, which can be seen as a peak in the literature distribution.

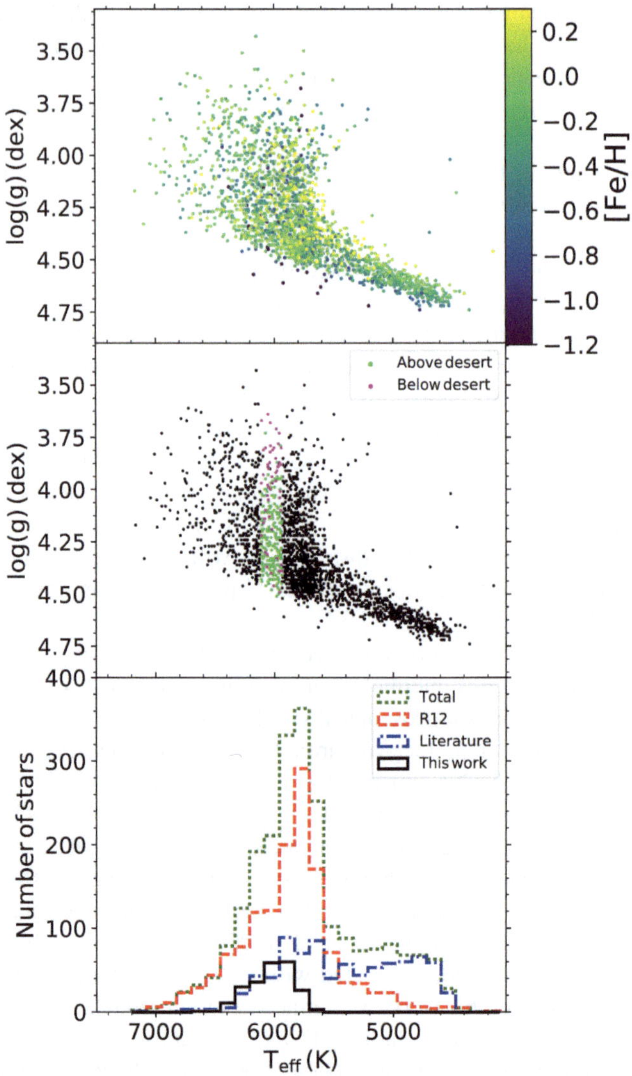

Fig. 6.1 Top panel and middle panel: HR diagram of stars in the final catalog. In the top panel, stars are color-coded by their metallicities and show that our sample contains dwarfs and subgiants. In the middle panel, we color-code only stars in the temperature range of the desert by their Li abundances. Bottom panel: Temperature distribution of stars in the final catalog and other sources, to see how they contribute to the catalog. New observations greatly increase the sample size in the temperature range of the desert. Credit: Aguilera-Gómez et al. 2018, A&A, 614, A55, reproduced with permission ©ESO

To obtain nonlocal thermodynamic equilibrium (NLTE) corrections, the microturbulence velocity is necessary. We estimate it by using the relations of Bruntt et al. (2012) and our effective temperatures and $\log g$. With this information and the grid of NLTE corrections by Lind et al. (2009a), we can only calculate these corrections for1285 of the stars in the sample, which are always smaller than $\Delta A(Li)_{NLTE} = 0.05$. With the smaller sample of NLTE Li abundances, we repeated the analysis presented in this chapter, without finding any differences in the correlations. We thus prefer to work with the LTE abundances, which provide a much larger sample to study.

A final detail for our catalog to be complete is the inclusion of a flag that refers to the planet-hosting status of the stars. We separate the sample into known planet hosts, stars with no known planets after radial velocity measurements, or unknown. This information is obtained from The Exoplanet Orbit Database (Han et al. 2014), Fischer and Valenti (2005), Sousa et al. (2008), Ghezzi et al. (2010), Brugamyer et al. (2011), and Delgado Mena et al. (2014, 2015). These works indicate that 225 are planet hosts, 1245 are not, and the remaining stars have unknown status.

The final catalog, with masses, ages, atmospheric parameters, Li abundances, and planet flag is available in electronic form at the CDS via anonymous ftp to cdsarc.u-strasbg.fr (130.79.128.5) or via http://cdsweb.u-strasbg.fr/cgi-bin/qcat?J/A+A/.

6.3 An Overview of the Catalog

The Li abundance and its trends with $\log g$, [Fe/H], mass, and age are found in Fig. 6.2, from top to bottom. Trends with $T_{\rm eff}$ will be discussed in Sect. 6.4. Unreliable ages can easily contaminate trends; that is why in the bottom panel, we only use reliable ages. The colors in the figure indicate the planet-hosting status, with green being planets hosts. Also, darker colors indicate that the Li is detected, while the lighter colors are Li upper limits.

We analyze trends found in each of the panels of Fig. 6.2. No relevant trends are distinguished with the $\log g$ of stars. Most of the stars in the sample have metallicities between [Fe/H] = − 0.5 and [Fe/H] = 0.5 although we do find some outliers at lower metallicities. Some of the known trends of planet hosts with metallicity (e.g. Gonzalez 1997; Fischer and Valenti 2005; Bond et al. 2006, among others) are recovered in this panel, with planet hosts being more metal-rich. However, this trend is not found for hosts of Neptune-like or Earth-like planets (Sousa et al. 2008). The stars in the sample tend to have lower masses, between 0.8 and 1.2 $M_\odot$. The positive correlation of Li abundances with mass can be explained by the nature of these stars. As most of them are dwarfs, we expect them to have shallower convective envelopes at larger masses, which provides the star less mass to dilute the Li. Part of the spread in this relation is given because trends of convection with Li are also metallicity dependent. We show how the metallicity affects this distribution in Fig. 6.3, with the same diagram as in the third panel of Fig. 6.2 now color-coded by the metallicity.

Finally, in the panel with ages, we see a decreasing correlation of the upper envelope of the distribution but with a very large spread. At younger ages, this correlation is even clearer.

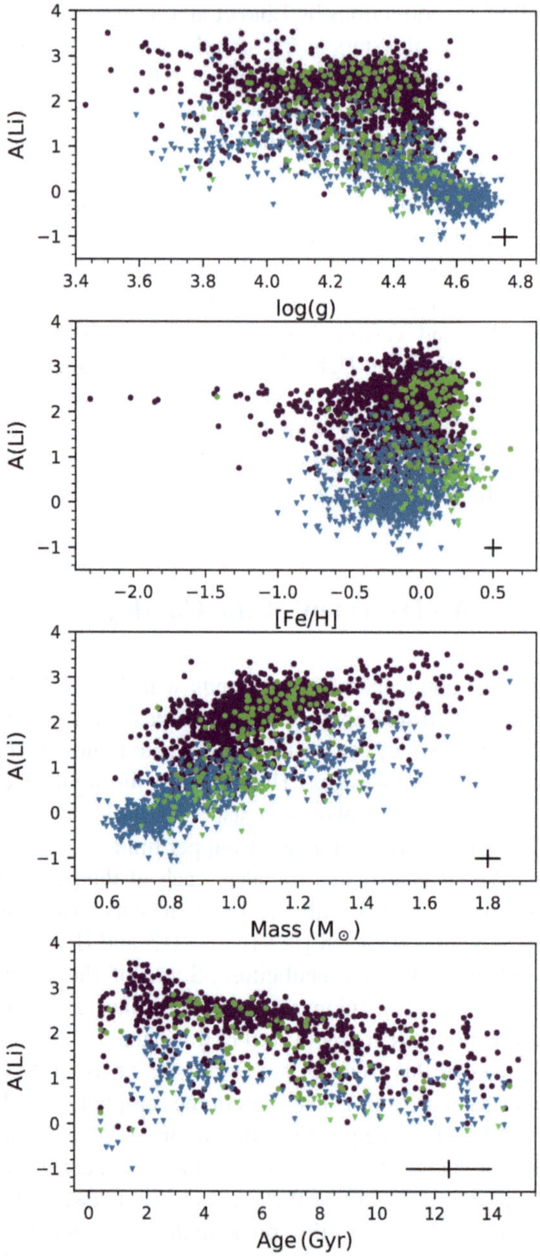

Fig. 6.2 Trends of Li abundance in our catalog stars with log g, [Fe/H], mass, and age (from top to bottom panels). Circles are Li determinations, while triangles are upper limits. Green symbols are planet hosts, and blue symbols are stars with no known planets so far. In the bottom panel, we only include stars with reliable ages. Credit: Aguilera-Gómez et al. 2018, A&A, 614, A55, reproduced with permission ©ESO

Solar twins are stars with similar characteristics to our Sun, where we should expect a decreasing correlation of Li with age. We restrict our sample to $T_{\rm eff} = T_{\rm eff,\odot} \pm 90$ K, $\log g = \log g_\odot \pm 0.2$, [Fe/H]=[Fe/H]$_\odot \pm 0.11$, and $M = M_\odot \pm 0.05$) and show in Fig. 6.4 that this correlation is found, although with important

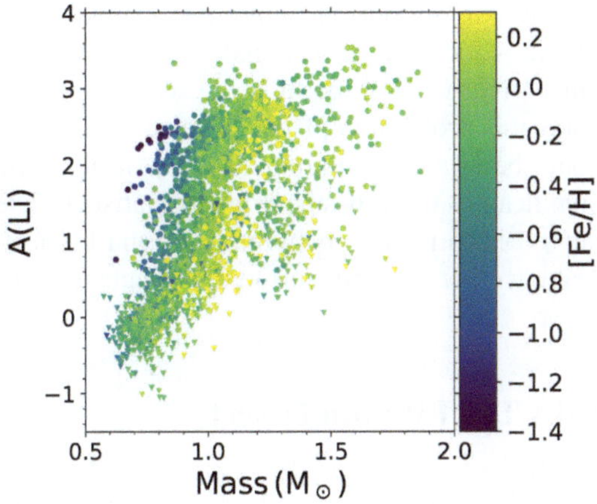

Fig. 6.3 Li abundances vs. mass. The color map shows the metallicities of the stars. Circles are Li determinations, while triangles are upper limits. Credit: Aguilera-Gómez et al. 2018, A&A, 614, A55, reproduced with permission ©ESO

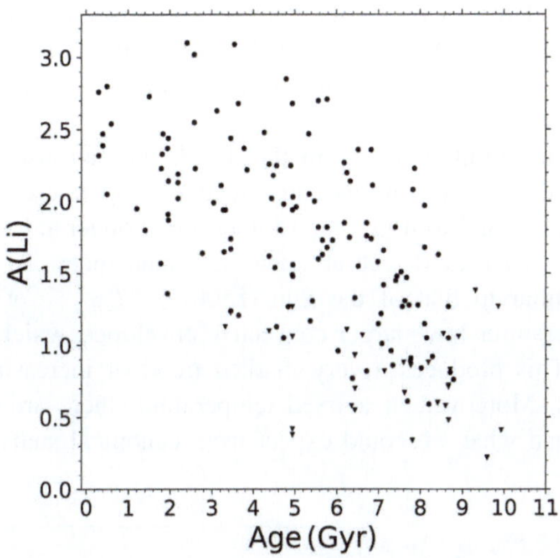

Fig. 6.4 Li abundance as a function of age for solar twins in our sample, which show a decreasing correlation but with considerable scatter. These stars are selected by ranges in effective temperature, log g, metallicity, and mass ($T_{\rm eff} = T_{{\rm eff},\odot} \pm 90$ K, $\log g = \log g_\odot \pm 0.2$, [Fe/H]=[Fe/H]$_\odot \pm 0.11$, and $M = M_\odot \pm 0.05$). Here, only reliable ages are considered. The points are Li detections, and the triangles represent upper limits. Credit: Aguilera-Gómez et al. 2018, A&A, 614, A55, reproduced with permission ©ESO

scatter. That we are finding stars with different Li abundances at a fixed age could be due to our large errors in this last parameter, but it could also be due to their different rotation rates (Baumann et al. 2010). Also studying solar twins inside the same limited ranges of parameters, Carlos et al. (2016) show a very tight correlation between Li and age; however, their errors are much smaller than ours.

At last, the relation between planet-hosting status and Li abundance has been a much-debated topic. In general, we do not see such a correlation here, partly because we are not restricting our sample to solar analogs and twins as do some of the works that have found a trend. We will not investigate this matter with our current catalog, because we would need to apply several cuts to our data to study this problem in planet hosts and nonhosts so they are homogeneous, and we could be artificially introducing bias in the different samples. This would reduce significantly the robustness of the statistics. There is no way to know for sure if any other correlation between hosts and Li or other parameters could be found by making other arbitrary cuts to the sample (e.g., Baumann et al. 2010).

6.4 The Lithium Desert

The lithium desert can still be identified in Fig. 6.5, in the region between 5950 and 6100 K as a relative lack of stars compared to adjacent regions. The polygon here is only indicative of the position of this desert and is not expected to show exactly where there are fewer stars. It is possible that the desert extends to higher temperatures as well.

Additional to the lithium desert, in this figure we can also see the possitive correlation between Li upper limits and effective temperature. This is produced because at constant signal-to-noise, the Li lines are stronger at lower temperatures, making the line easier to detect at lower Li abundances. At higher effective temperatures, similar to that of the Sun ($5500 \lesssim T_{\rm eff} \lesssim 6000$ K), the stars with lower temperatures have larger convective envelopes, which leads to smaller Li abundances. This produces a very marked trend of increasing Li abundance with temperature. Moreover, at a fixed temperature, there are stars with lower Li abundances than what we could expect from canonical stellar evolution (e.g.,

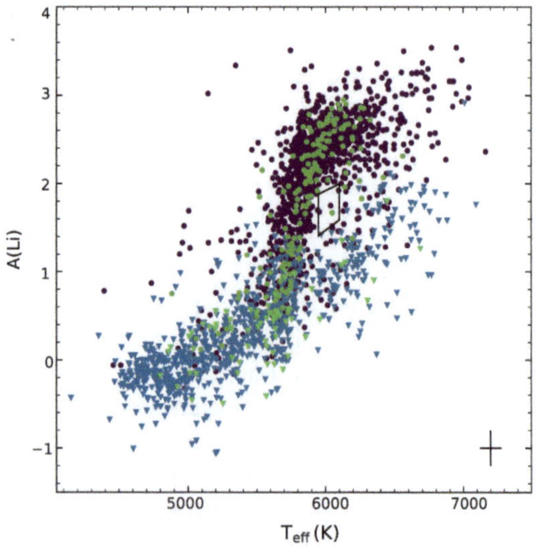

Fig. 6.5 Li abundance vs. effective temperature. Symbols are as in Fig. 6.2. The polygon shows the approximate position of the Li desert, as suggested by R12. Credit: Aguilera-Gómez et al. 2018, A&A, 614, A55, reproduced with permission ©ESO

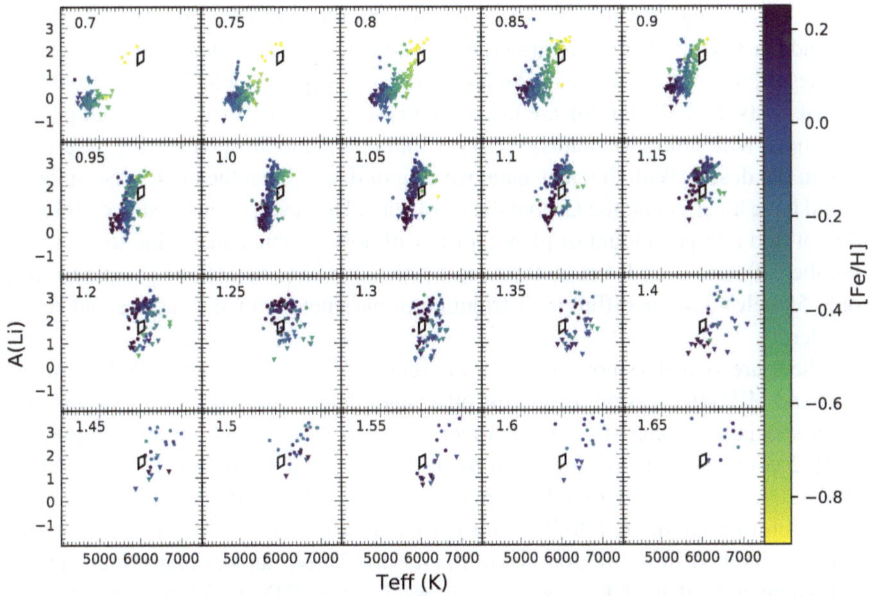

Fig. 6.6 Li abundance vs. T_{eff} for all stars in the sample, subdivided into bins of mass ($\mathrm{M}\pm0.025\,\mathrm{M}_\odot$). The circles are Li determinations, triangles are upper limits, and colors indicate the stars' metallicity. Credit: Aguilera-Gómez et al. 2018, A&A, 614, A55, reproduced with permission ©ESO

Pinsonneault 1997), which indicates that there is a depletion mechanism acting on the main sequence, decreasing the Li.

The feature of the Li desert is still not understood completely although some works have previously attempted to find the nature of this region. R12, finding that at a specific mass there are stars with both low and high Li abundances (e.g., Fig. 6.6, panel $1.1\,\mathrm{M}_\odot$), suggest that a mechanism of Li depletion might be acting as the star evolves to the subgiant branch, making stars move in the Li abundance vs. effective temperature diagram from the higher Li abundance side of the desert to the lower side. If there is a depletion mechanism acting on stars, then it should be dramatic, as almost no stars are found in the region of the desert. On a different alternative, Chen et al. (2001) suggest that the low Li abundance side of the desert could be related to the Li dip.

With our compiled catalog, we study the origin of the Li desert. This feature is still present in Fig. 6.5 but less clear than in previous works. With the increased sample size, we now find a few stars in a region that was previously completely empty. However, now doubling the amount of stars, this region still seems to have much less stars than adjacent zones. Another interesting point about the desert is that the lower side is populated by upper limits, which means that the desert could extend towards lower Li abundances. We are not able to determine precisely how much Li do these stars really have, and they may be completely depleted.

R12 also suggested that the Li desert may be related to the presence of planets around stars, with planets having higher Li abundances in the temperature range of the desert, all located above it. We analyze this in Fig. 6.5, where green points are planet hosts, and blue points are the comparison sample (both stars without planets and those with unknown status). We now find one planet host below the desert and one in the desert. With this new data, we cannot discard that the presence of planets could have an effect on the Li abundance of stars. However, in this case we probably do not find a large amount of planet hosts with low Li abundances due to the lower number of stars overall below the desert (there are 251 stars above the desert, and only 55 below) or the different distribution of parameters for stars above and below the desert.

There are two stars now found in the region of the desert, HIP22826, a planet host and HIP51028. Analyzing their other parameters, these stars do not have any other special characteristics. Three other stars that were already in the compilation of R12 could also be considered to be in the region of the desert. López-Valdivia et al. (2015) had already found before one star in the desert, BD+47 3218, and other in the limiting region, HIP30552. This particular star is also included in our catalog, and it is located in the region above the desert. These authors claim their data is consistent with that of R12, so we can assume that BD+47 3218 is still located inside the region of the desert. Although there are some stars in the Li desert, the region with less stars does not seems to be just a product of selection effects or uncertainties in the measurements.

In this section, we analyze correlations of parameters in the regions around the desert, and based on how these parameters affect the Li abundance, we find the origin of the Li desert.

6.4.1 Characterizing Stars Around the Li Desert

Using our extended catalog, we study how A(Li) changes with stellar parameters and identify stars that surround the region of the Li desert. The feature of the desert is found in the A(Li)-$T_{\rm eff}$ plane. We look for correlations in this same plane by dividing the sample into smaller mass ranges and color-coding it by the metallicity in Fig. 6.6. Once again, the position of the desert is marked by the polygon.

The panels of Fig. 6.6 show that the temperatures of the stars increase with mass, a known effect in main sequence stars. We also see a metallicity gradient in each of the panels, with metal-rich stars being cooler. Here we recover the trend between mass and Li abundance of stars seen in the third panel of Fig. 6.2, with Li increasing with mass. Moreover, the metal-rich stars tend to be more Li-depleted than their metal-poor counterparts. This can be explained solely by the size of the convective zone, given that for the same mass, stars with higher metallicity have larger opacity, making their temperatures lower and surface convection more important, diluting the Li. These trends with metallicity fade for stars of mass higher than $1.2\,M_{\odot}$.

We see that stars with masses $M < 0.8\,\mathrm{M}_{\odot}$ are not surrounding the region of the desert and as such do not contribute to its formation. There seems to be a difference in the mass of stars populating the desert. The upper side of the desert is populated by stars with masses from 0.8 to $1.35\,\mathrm{M}_{\odot}$, while the lower side of the desert is populated by more massive stars, with masses from $\sim$ 1.1 to $1.55\,\mathrm{M}_{\odot}$.

The lower side also seems to be populated by more metal-rich stars compared to the upper side, populated by stars in a wide range of metallicities. At a specific range of masses (around one solar mass), more metal-rich stars are more Li-depleted, so there could be an effect of mass behind the different metallicity distributions above and below the desert. For higher mass stars, as those located in the lower side of the desert, there is no clear correlation between metallicity and Li abundance. These stars have shallower convective envelopes, and even higher mass objects during the main sequence have convective cores and radiative interiors, being this the most likely explanation for the lack of metallicity correlation.

The different distributions of parameters for stars surrounding the Li desert indicate that we have to disentangle the effects of mass and metallicity on Li abundance before concluding about the origin of this feature. It seems there is a considerable effect of mass, but the presence of stars with high and low Li abundances with the same mass indicates that there may be something more in the formation of the desert.

6.4.2 *Characterizing the Mass, Metallicity, and Evolutionary Stage of the Li Desert*

To more closely characterize the Li desert, we have studied a more precise indicator of the evolutionary stage for each of the stars in the T_{eff} range of the desert, by using stellar models. The Li abundance has a complicated behavior depending on several stellar parameters, such as temperature, age, metallicity, and mass. When using these same parameters to find trends and conclude, we could be introducing biases in the relations. Then, we use as a better indication of evolutionary phase the hydrogen fraction in the core of the star.

This fraction is calculated from stellar models obtained with the Yale Rotating Evolutionary Code (YREC, Demarque et al. 2008). Each star already has associated a mass and metallicity that we use as initial conditions for the models. Then, we identify when the T_{eff} and the absolute magnitude of the model are the same as those that we have for the star. We prefer to identify the correct step in the models using these two parameters instead of age, as ages by isochrone fitting can be very uncertain.

Models also have information on the composition of the stellar core. Once the correct model and timestep are found, we obtain the hydrogen fraction in the core, information used to produce the color-coded HR diagram in Fig. 6.7. In the left panel, we see the stars above the Li desert, while in the right panel, stars below

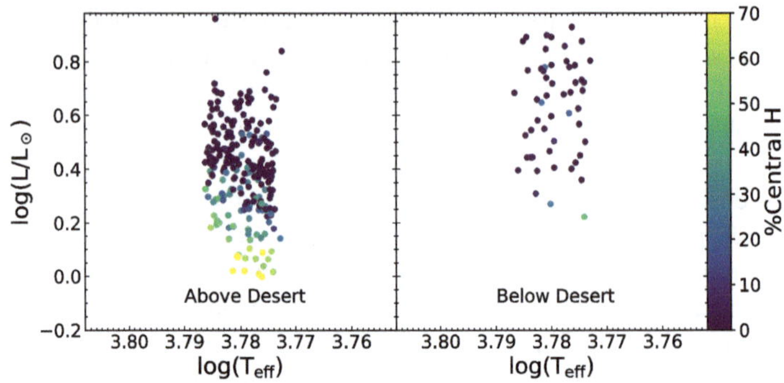

Fig. 6.7 HR diagram of stars above (left panel) and below (right panel) the Li desert. Colors indicate the percentage of hydrogen in the core of stars, which show that those stars below the desert tend to have less core hydrogen and thus that are more evolved than those stars located above the desert. Stars below the desert are also more luminous, indicating that there could be a mass effect that we should consider. Credit: Aguilera-Gómez et al. 2018, A&A, 614, A55, reproduced with permission ©ESO

the Li desert are depicted. The color of the points indicates the percentage of central hydrogen that allows us to clearly identify stars still in the main sequence phase from those that do not have hydrogen in their core (dark blue symbols). Stars below the desert seem to have a lower core hydrogen fraction, being closer to evolved stages of evolution; also, they are more luminous overall. The different luminosity of stars below the desert could also be an effect of their higher average masses. Being stars of higher mass, they have shorter main sequence lifetimes, which mean they evolve more rapidly, having a lower percentage of core hydrogen.

The distributions for mass, metallicity, age, and core hydrogen, for stars above and below the desert, can be seen in Fig. 6.8 from left to right. Top panels show the frequency, and bottom panels show the cumulative distribution. As suggested previously, the mass distribution shows that stars below the desert tend to be more massive than those above. The distribution of metallicity is similar for stars above and below the desert. The age is also very similar, but stars with higher Li abundances in this region are somewhat older. The distribution of core hydrogen percentage is different for stars above and below the Li desert. As we saw in Fig. 6.7, the lower side of the desert is more evolved, having stars above the desert a higher hydrogen fraction. The different distribution between stars above and below the desert is mainly given by the number of stars in the first bin, with hydrogen percentage close to 0%, but also, it is reinforced by the lack of stars with more than 30% core hydrogen in the lower side of the desert.

Stars below the desert are more evolved and have already left the main sequence. To find the origin of the desert, we need to find where do these stars come from, for which we need to isolate the effect of different stellar parameters on Li.

From the difference in the two distributions in each of these parameters, a possibility is that there is a depletion mechanism acting inside stars when they evolve to subgiants, thus making the Li decrease considerably or even completely.

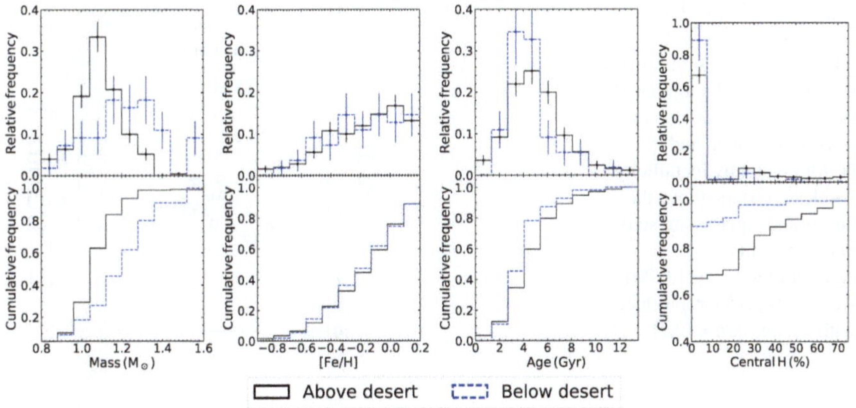

Fig. 6.8 Distribution of mass, metallicity, age, and hydrogen fraction in the core from left to right. Different colors and linestyles represent stars above and below the Li desert (high and low Li abundances respectively). Top panels shown frequency, while lower panels show cumulative distribution. Credit: Aguilera-Gómez et al. 2018, A&A, 614, A55, reproduced with permission ©ESO

This is consistent with what was previously suggested by R12. But, it is also possible that these subgiants are simply evolving from MS stars that originally had less Li, for example, those from the Li dip.

6.4.3 The Origin of the Li Desert

Abnormal stars in the effective temperature range of the Li desert are those with low Li abundances because with these temperatures stars are predicted by evolutionary models to have higher Li abundances. Also, as we found previously, the low Li stars are also more evolved. For the purpose of identifying the origin of the Li desert, we try to find the progenitors of these stars. There are two possibilities that we study. The first is that they have evolved from stars above the Li desert, which, with a very efficient mechanism of Li depletion, rapidly traverse through the region of the desert (As suggested by R12). The second possibility is that these stars below the desert are evolving from stars that already had very low Li abundance in the main sequence, such as those from the Li dip. To make conclusions regarding the origin of the Li desert, we combine information on the evolutionary status of stars and their properties such as mass and metallicity.

Stars Above the Desert and Stars Below the Desert

Is it possible that stars below the desert are evolving from those above it? We find in this section that atmospheric parameters of stars above and below the desert

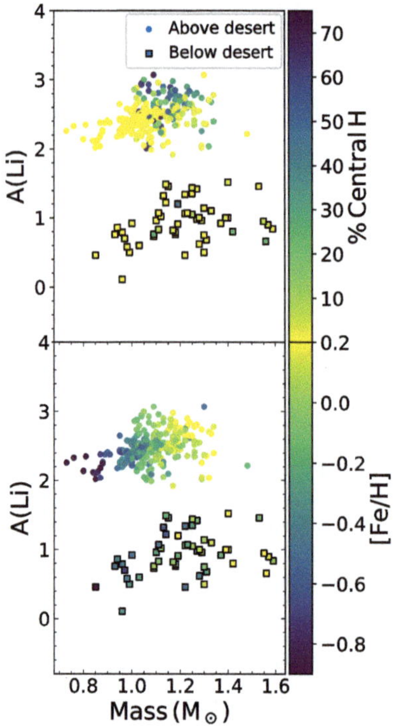

Fig. 6.9 Lithium abundances vs. mass for stars in the effective temperature range of the Li desert. The top panel is color-coded by the percentage of core hydrogen and the bottom panel by metallicity. Circles are stars above the desert and squares are stars below. Credit: Aguilera-Gómez et al. 2018, A&A, 614, A55, reproduced with permission ©ESO

cover a similar range. However, the distributions are different. This implies that the progenitors of the stars below the desert are not located in this effective temperature range.

Figure 6.9 shows the lithium abundance as a function of mass for stars located in the effective temperature range of the desert. The top panel is color-coded by the evolutionary status of the stars (percentage of core hydrogen), and the bottom panel, by metallicity. The top panel reinforces that stars below the desert are more evolved (lighter colors). At a certain fixed mass, we can find stars with low and high Li abundances, so mass cannot be the only parameter contributing to the formation of the desert. Focusing only on one mass, we see that stars below the desert are more evolved than those above.

If stars above the desert evolve to become stars below the desert, decreasing somehow their Li abundances, we must find main sequence stars with high Li abundances that have the same mass and metallicity than stars below the desert. Stars of the same mass and metallicity are expected to go through the same evolution (without considering mixing or other parameters that could modify the Li abundance, for example, the initial rotation rate).

We expect stars of the same mass and metallicity to follow similar evolutionary tracks and thus have similar Li abundances. We can track stars with the same atmospheric properties as they get older to see how the Li evolves in them. Stars eventually reach the subgiant branch, where little to no change is expected in the convective envelope, so stars should preserve the Li abundance they had at the end of their main sequence phase.

But we see in Fig. 6.9 that only considering stars above and below the Li desert limits the mass of stars in the main sequence to a very specific range. Stars more massive than M> $1.3\,M_\odot$ and less massive than M< $1.0\,M_\odot$ are only subgiants, and we find no main sequence stars in these mass ranges. It is not so strange to find only subgiants here, because stars in the region of the desert are only selected by their temperatures. The main sequence stars of higher mass also have higher temperatures that fall outside the range of temperatures of the desert. Also, although the temperature of lower-mass stars depends on the metallicity, they are colder, usually not reaching the high temperatures of the Li desert.

We confirm that stars above and below the desert have intrinsically different properties by analyzing the bottom panel of Fig. 6.9, same as the top panel but color-coded by metallicity. Below the desert, all stars have [Fe/H]>−0.45, but more importantly, if we focus on a specific mass range, stars with less Li abundance are always more metal-poor than those above, showing darker colors in the figure. Figure 6.10 shows exactly this point with the metallicity distribution of stars. We only plot two limited mass ranges. In the left panel, the mass is $1.10 \pm 0.05\,M_\odot$, and in the right panel, it is $1.20 \pm 0.05\,M_\odot$. The distribution of stars above the desert at both ranges of masses peaks at higher metallicity, implying that they are more metal-rich.

In conclusion, as stars of the same mass have different metallicities, they do not follow the same evolution and can have very different Li abundances. We should not compare stars above with those below the desert. They do not come from the same set of stars with similar properties and could have different Li abundances initially. The comparison we should attempt is between stars below the desert and main sequence stars with the same mass and metallicities, which may or may not fall in the temperature range of the Li desert.

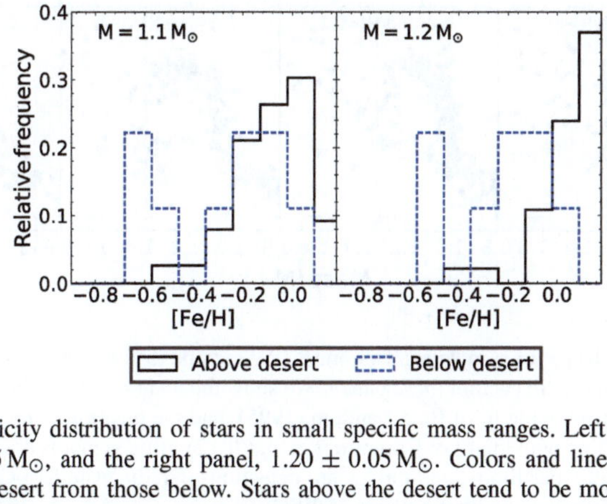

Fig. 6.10 Metallicity distribution of stars in small specific mass ranges. Left panel shows stars with $1.10 \pm 0.05\,M_\odot$, and the right panel, $1.20 \pm 0.05\,M_\odot$. Colors and line styles distinguish stars above the desert from those below. Stars above the desert tend to be more metal-rich than stars below at a fixed mass. Credit: Aguilera-Gómez et al. 2018, A&A, 614, A55, reproduced with permission ©ESO

Stars Below the Desert and Stars in the Li Dip

Given that stars below the desert do not evolve from those above, could they evolve from the Li dip, main sequence stars that originally had much less Li? To explore this option, we use the full homogeneous catalog, because progenitors of low Li abundance stars in the desert are not in that effective temperature range. We conclude that stars below the desert are consistent with having evolved from the Li dip.

The different metallicity bins in Fig. 6.11 allow to identify the Li dip. This feature was first discovered in populations with restricted metallicities. Thus, we create something similar to clusters with field stars, restricting ranges of metallicities. In this diagram, the Li dip should be seen as a decrease in Li abundance at a certain mass. For clusters the signal is clear, but it is harder to identify with field stars because the signal is not a sudden drop in Li abundance, but an increase in the number of upper limits.

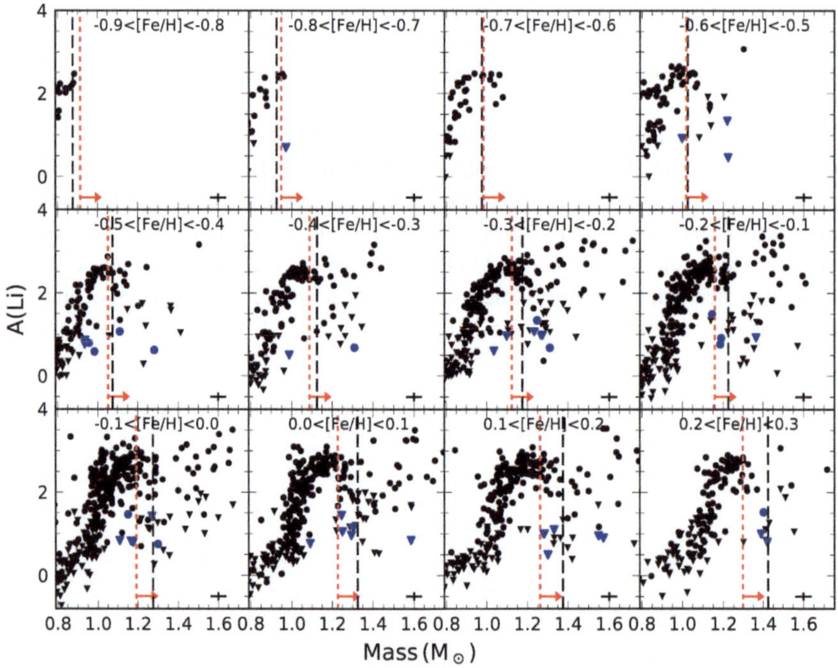

Fig. 6.11 Li abundance for sample stars vs. mass. We include different metallicity bins to clarify the feature of the Li dip. In each of these panels, we show the location of the center of the Li dip as a dashed black line, with the fit of Balachandran (1995), and the position of the cooler side of the dip as a red dotted line with the fit of Cummings et al. (2012). Blue symbols are stars below the Li desert, triangles are upper limits, and circles are detections. Credit: Aguilera-Gómez et al. 2018, A&A, 614, A55, reproduced with permission ©ESO

When focusing on each of these panels, more massive stars have higher Li abundances, with a clear correlation for almost all metallicity bins, at least until stars reach a particular mass where the Li decreases (As an example, in the bin $-0.4 <$[Fe/H]< -0.3, the drop is located between masses ~ 1.1 and $1.3\,M_\odot$). This is the Li dip. The exact mass where the drop (or spread) is found depends on the metallicity, and what we find here is consistent with what is found in other works both in clusters (Balachandran 1995; Anthony-Twarog et al. 2009; Cummings et al. 2012; François et al. 2013) and in the field (e.g., Lambert and Reddy 2004; Delgado Mena et al. 2015).

The position of the Li dip is indicated in each panel of Fig. 6.11, the center of the Li dip as a dashed black line, with the fit of Balachandran (1995), and the position of the cooler side of the dip as a red dotted line with the fit of Cummings et al. (2012). The first fit is $M/M_\odot = 1.3 + 0.5$[Fe/H], and the second fit is $M/M_\odot = 1.21 + 0.35$[Fe/H]. The arrow shows the direction where we find the dip. The center and hot side of the Li dip can sometimes be difficult to identify, especially when older clusters are being studied and stars at the hot side of the dip already evolved. The position of the dip given by these two fits is consistent with the location of depleted stars that we find in this work, independent of their evolutionary status.

Stars below the Li desert are in blue symbols in Fig. 6.11. We can compare these with other stars of similar parameters. The position of stars below the desert in this diagram is not exclusive, and these stars are usually surrounded by much others of the catalog. In conclusion, stars with low Li in the region of the desert are not unusual, and other stars in the sample with the same mass and metallicity also have low Li abundances. Then, these stars do not require extra-mixing or additional depletion to be explained, and it is possible that main sequence stars with low Li abundances (located at a similar position in this diagrams with mass and metallicity) will evolve into effective temperatures typical of the desert.

Furthermore, stars below the Li desert have positions consistent with the Li dip and could have evolved from those stars. There are some exceptions to this that we analyze. With a metallicity $-0.5 <$[Fe/H]< -0.4, there are four stars with extremely low masses, not sharing a position with the dip. However, for those masses, Li is expected to be much higher. Within uncertainties for these objects, they could still be consistent with $\sim 0.9\,M_\odot$, and thus their Li do not require any other mixing mechanism or further explanation. This is different to the case of the dwarf HIP22589 with near-solar metallicity, HIP50805 with [Fe/H]$= -0.26$, and HIP49988, with [Fe/H]$= -0.5$, three stars that are far away from any other star in the diagrams and seem to be especially Li-depleted. Both HIP49988 (Ženovienė et al. 2015) and HIP50805 are stars enriched in s-process elements, and HIP50805 is a CH subgiant (Luck and Bond 1982), so they are chemically peculiar stars that could be produced by mass transfer, which would also change their Li abundance. HIP22589 is a main sequence star in the lower side of the desert, which makes this a very peculiar case anyway.

All other stars with low Li abundances in the temperature range of the desert are located in a similar position to other stars with similar lithium abundances. If there are already depleted main sequence stars in similar positions, the subgiants below the desert could just follow canonical stellar evolution from those progenitors. In the higher metallicity bins, there is no longer main sequence star that we can compare to, but the position of the dip with metallicity is consistent, and stars in the desert at those metallicities could still be evolved from the dip.

Based on what we found regarding the origin of the desert, the exact position of this feature and what we marked with the polygon are of little importance, as the desert is finally just an artifact. The position of these stars in the A(Li)-T_{eff} below where the desert was found is coincidental. Stars with low Li are not special, and they do not require extra-mixing in the subgiant phase. The low Li abundances they had in the main sequence (Li dip) are enough to explain them. We can only find these stars because as they evolve, they move to a particular region in this plane where we are used to finding only stars with higher Li abundances.

6.5 Discussion

We conclude that stars with low Li abundances in the temperature range 5950 to 6100 K are consistent with other stars of similar parameters and could have evolved from the Li dip. We do not need any extra Li depletion mechanism to explain the low Li of these stars. In this section, we discuss previous works on the subject of the Li desert and the Li dip.

Balachandran (1990) presented a sample of Li-depleted stars with masses similar to the Li dip, and they suggest that these stars could be field analogs to the dip. These Li poor stars have evolved from the main sequence stars with higher temperatures between 6500 and 6800. The Li dip is a possible origin for these stars. However, they are not found in the temperature of the Li desert, so they are not the same stars we study here, although the situation could be somehow equivalent.

The feature of the Li desert has become clearer in years as we improve the Li upper limits, but Lambert et al. (1991) could already see that stars with temperatures higher than 5900 K show a bimodal distribution in lithium, with some of the stars having determinations, and some others with only upper limits. They claim that the low Li abundance side could have evolved from the Li dip. The temperature range of the studied stars is, again, higher than what we study in this work.

In another work, Chen et al. (2001) also identify the feature of the Li desert, with $T_{\mathrm{eff}} > 5900$ K. Here a gap is already visible between both sides of the desert. As these stars have very specific masses and they follow a mass-metallicity trend similar to that suggested for the Li dip (Balachandran 1995), the authors suggest that the dip and the Li-depleted stars below the desert could be related. R12 showed that this relationship between mass and metallicity is observed in all the stars, both Li-normal and Li-poor, and thus although the conclusion is similar to what we find in this work with a completely different method, we do not think the mass-metallicity dependence of stars is enough to arrive at that conclusion accurately.

With this work, we also reinforce the conclusion that the effective temperature is not enough to identify stars that belong to the Li dip or have evolved from it. Mass and, as we could see in this work, metallicity are both important when Li depletion is involved.

6.6 Summary

The Li abundance can be useful to study several different topics in astrophysics. However, as it has a complicated dependency on stellar parameters, a large homogeneous catalog is needed, so that conclusions are based on large samples of stars with no internal systematic differences.

We have measured atmospheric parameters and Li abundances for a sample of 227 stars, with effective temperatures close to the region of the Li desert. We then create a large catalog of stars by combining previous literature samples as Delgado Mena et al. (2014, 2015) and Gonzalez (2014, 2015) homogenized to the same scale. After cleaning the sample of binaries, we only consider stars with known parallaxes to find masses and ages. The final catalog has 2318 dwarfs and subgiants.

The region in effective temperature between solar-type stars and the Li dip (between 5950 and 6100 K) has not received much attention. Previous works have found that the behavior of Li in stars in this temperature range is peculiar, as they seem to separate discontinuously into stars with only Li upper limits and those with detections. That region in the A(Li)-$T_{\rm eff}$ with significantly less number of stars is the Li desert. At those effective temperatures, canonical stellar evolution predicts stars with high Li abundances. Thus, the stars with depleted Li are unusual. We attempt to find an explanation for those stars.

Also, Li-depleted stars in the main sequence are ideal progenitors of RGB stars with low Li abundances, that allow to easily identify Li-enriched giants and study associated processes, such as substellar companion engulfment and internal mixing.

Different explanations have been suggested for the lack of stars in the region of the Li desert. One of these suggestions is that the low side of the desert is associated with the Li dip, based on the mass-metallicity dependence found for stars below the desert, a correlation that was previously found in the Li dip. R12 recently question this association, because all stars, regardless of their Li abundance show this mass-metallicity correlation. They propose some alternative explanations. It is possible that the desert is associated with the presence of planets around stars, as they find that planet hosts are located above the desert. On the other hand, it is also possible that a Li depletion mechanism is acting on stars above the desert, making stars rapidly cross the diagram of A(Li)-$T_{\rm eff}$ to lower Li abundances.

With the compiled homogeneous catalog, we find the origin of the Li desert, combining the stellar properties with stellar evolutionary models. Our catalog also includes information about the planet-hosting status. We cannot discard with the extended catalog that there is an effect of planets in the Li abundance of stars, but the trend found by R12 could also be related to the number of stars above and below the desert and not to a real physical effect.

Calculating the evolutionary stage of stars we find that the majority of stars with low Li abundance in the region of the desert have already evolved off the main sequence. To identify the progenitors of these stars, we find correlations of Li abundance with other parameters. In general, we see that stars above and below the desert have different masses, with stars of high Li abundances being less massive.

We recover some expected correlations, such as that metal-rich stars of the same mass are more Li-depleted due to their larger opacities and correspondingly more important convection, although the correlation is less clear for higher masses. Also, older stars do not contribute to the formation of the Li desert.

We have analyzed stars in the temperature range of the desert finding that there are important differences in the distribution of parameters. At a given mass, stars below the desert are more metal-poor than stars of that same mass but with higher Li abundances. Thus, we are unable to find the progenitors of stars below the desert, above it. Stars above the desert do not evolve rapidly depleting their Li abundances.

Since we are only selecting stars in a very specific temperature range, and the distribution of stellar parameters of stars above and below the Li desert are different, only analyzing this sample does not allow to correctly identify the origin of the Li desert.

We use the entire catalog of Li abundances to find the progenitors of Li-depleted stars in the region of the desert. In particular, we identify the Li dip using these fields stars. This feature is much harder to find in field stars than clusters, but we attempt to recreate the conditions of clusters here, by grouping stars in restricted ranges of metallicity. Although the dip is not seen as a drop in the Li abundance at a given mass for each metallicity, there is a considerable spread in Li and an increase in the number of upper limits. That is the feature of the Li dip. We recover stars from the Li dip and how the masses of the Li dip change with metallicity (Balachandran 1995).

The evolutionary path followed by stars depends on their masses and metallicities. Thus, we expect stars with similar parameters to have similar evolutions. In consequence, we can identify stars with the same mass and metallicity than subgiants below the Li desert, and they usually are located in the region of the Li dip. Then, stars below the desert are consistent with being evolved from Li dip stars. Although we can link stars from the desert with those from the dip, the physical mechanism producing the depletion in the Li dip is still unknown.

The future of these and other studies requiring masses of stars is in the *Gaia* Mission, combined with spectroscopic surveys or follow-up with high-resolution spectrographs. With this, we could confirm the origin of stars in the Li desert and also find the mechanism behind the Li depletion for stars in the Li dip. Additionally, this would allow to find interesting samples to study other Li-related problems, like the enrichment of giants.

References

C. Aguilera-Gómez, I. Ramírez, J. Chanamé, A&A **614**, A55 (2018). http://doi.org/10.1051/0004-6361/201732209

B.J. Anthony-Twarog, C.P. Deliyannis, B.A. Twarog, K.V. Croxall, J.D. Cummings, AJ **138**, 1171 (2009). http://doi.org/10.1088/0004-6256/138/4/1171

S. Balachandran, ApJ **354**, 310 (1990). http://doi.org/10.1086/168691

S. Balachandran, ApJ **446**, 203 (1995). http://doi.org/10.1086/175779

S.A. Barnes, ApJ **586**, 464 (2003). http://doi.org/10.1086/367639
S.A. Barnes, ApJ **669**, 1167 (2007). http://doi.org/10.1086/519295
P. Baumann, I. Ramírez, J. Meléndez, M. Asplund, K. Lind, A&A **519**, A87 (2010). http://doi.org/10.1051/0004-6361/201015137
J.C. Bond, C.G. Tinney, R.P. Butler, H.R.A. Jones, G.W. Marcy, A.J. Penny, B.D. Carter, MNRAS **370**, 163 (2006). http://doi.org/10.1111/j.1365-2966.2006.10459.x
E. Brugamyer, S.E. Dodson-Robinson, W.D. Cochran, C. Sneden, ApJ **738**, 97 (2011). http://doi.org/10.1088/0004-637X/738/1/97
H. Bruntt, S. Basu, B. Smalley, et al., MNRAS **423**, 122 (2012). http://doi.org/10.1111/j.1365-2966.2012.20686.x
M. Carlos, P.E. Nissen, J. Meléndez, A&A **587**, A100 (2016). http://doi.org/10.1051/0004-6361/201527478
L. Casagrande, I. Ramírez, J. Meléndez, M. Bessell, M. Asplund, A&A **512**, A54 (2010). http://doi.org/10.1051/0004-6361/200913204
J. Chanamé, I. Ramírez, ApJ **746**, 102 (2012). http://doi.org/10.1088/0004-637X/746/1/102
Y.Q. Chen, P.E. Nissen, T. Benoni, G. Zhao, A&A **371**, 943 (2001). http://doi.org/10.1051/0004-6361:20010371
J.D. Cummings, C.P. Deliyannis, B. Anthony-Twarog, B. Twarog, R.M. Maderak, AJ **144**, 137 (2012). http://doi.org/10.1088/0004-6256/144/5/137
E. Delgado Mena, G. Israelian, J.I. González Hernández, et al., A&A **562**, A92 (2014). http://doi.org/10.1051/0004-6361/201321493
E. Delgado Mena, S. Bertrán de Lis, V.Z. Adibekyan, et al., A&A **576**, A69 (2015). http://doi.org/10.1051/0004-6361/201425433
P. Demarque, D.B. Guenther, L.H. Li, A. Mazumdar, C.W. Straka, Ap&SS **316**, 31 (2008). http://doi.org/10.1007/s10509-007-9698-y
D.A. Fischer, J. Valenti, ApJ **622**, 1102 (2005). http://doi.org/10.1086/428383
P. François, L. Pasquini, K. Biazzo, P. Bonifacio, R. Palsa, A&A **552**, A136 (2013). http://doi.org/10.1051/0004-6361/201220958
Gaia Collaboration, A.G.A. Brown, A. Vallenari, et al., A&A **595**, A2 (2016). http://doi.org/10.1051/0004-6361/201629512
E.J. Gaidos, G.W. Henry, S.M. Henry, AJ **120**, 1006 (2000). http://doi.org/10.1086/301488
L. Ghezzi, K. Cunha, V.V. Smith, R. de la Reza, ApJ **724**, 154 (2010). http://doi.org/10.1088/0004-637X/724/1/154
G. Gonzalez, MNRAS **285**, 403 (1997). http://doi.org/10.1093/mnras/285.2.403
G. Gonzalez, MNRAS **441**, 1201 (2014). http://doi.org/10.1093/mnras/stu661
G. Gonzalez, MNRAS **446**, 1020 (2015). http://doi.org/10.1093/mnras/stu2156
G. Gonzalez, M.K. Carlson, R.W. Tobin, MNRAS **403**, 1368 (2010). http://doi.org/10.1111/j.1365-2966.2009.16195.x
G. Guiglion, P. de Laverny, A. Recio-Blanco, et al., ArXiv e-prints: 1608.03411 (2016)
B. Gustafsson, B. Edvardsson, K. Eriksson, U.G. Jørgensen, Å. Nordlund, B. Plez, A&A **486**, 951 (2008). http://doi.org/10.1051/0004-6361:200809724
E. Han, S.X. Wang, J.T. Wright, Y.K. Feng, M. Zhao, O. Fakhouri, J.I. Brown, C. Hancock, PASP **126**, 827 (2014). http://doi.org/10.1086/678447
G. Israelian, E. Delgado Mena, N.C. Santos, et al., Nature **462**, 189 (2009). http://doi.org/10.1038/nature08483
D.L. Lambert, J.E. Heath, B. Edvardsson, MNRAS **253**, 610 (1991). http://doi.org/10.1093/mnras/253.4.610
D.L. Lambert, B.E. Reddy, MNRAS **349**, 757 (2004). http://doi.org/10.1111/j.1365-2966.2004.07557.x
K. Lind, M. Asplund, P.S. Barklem, A&A **503**, 541 (2009). http://doi.org/10.1051/0004-6361/200912221
R. López-Valdivia, J.B. Hernández-Águila, E. Bertone, et al., MNRAS **451**, 4368 (2015). http://doi.org/10.1093/mnras/stv1222
R.E. Luck, H.E. Bond, ApJ **259**, 792 (1982). http://doi.org/10.1086/160215

R.E. Luck, U. Heiter, AJ **131**, 3069 (2006). http://doi.org/10.1086/504080
E.E. Mamajek, L.A. Hillenbrand, ApJ **687**, 1264-1293 (2008). http://doi.org/10.1086/591785
M. Pinsonneault, ARA&A **35**, 557 (1997). http://doi.org/10.1146/annurev.astro.35.1.557
N. Pizzolato, A. Maggio, G. Micela, S. Sciortino, P. Ventura, A&A **397**, 147 (2003). http://doi.org/10.1051/0004-6361:20021560
D. Pourbaix, A.A. Tokovinin, A.H. Batten, et al., A&A **424**, 727 (2004). http://doi.org/10.1051/0004-6361:20041213
I. Ramírez, J.R. Fish, D.L. Lambert, C. Allende Prieto, ApJ **756**, 46 (2012). http://doi.org/10.1088/0004-637X/756/1/46
I. Ramírez, J. Meléndez, J. Bean, et al., A&A **572**, A48 (2014). http://doi.org/10.1051/0004-6361/201424244
B.E. Reddy, J. Tomkin, D.L. Lambert, C. Allende Prieto, MNRAS **340**, 304 (2003). http://doi.org/10.1046/j.1365-8711.2003.06305.x
C. Sneden, ApJ **184**, 839 (1973). http://doi.org/10.1086/152374
S.G. Sousa, N.C. Santos, M. Mayor, et al., A&A **487**, 373 (2008). http://doi.org/10.1051/0004-6361:200809698
K. Strassmeier, A. Washuettl, T. Granzer, M. Scheck, M. Weber, A&AS **142**, 275 (2000). http://doi.org/10.1051/aas:2000328
Y. Takeda, S. Honda, S. Kawanomoto, H. Ando, T. Sakurai, A&A **515**, A93 (2010). http://doi.org/10.1051/0004-6361/200913897
R. Ženovienė, G. Tautvaišienė, B. Nordström, E. Stonkutė, G. Barisevičius, A&A **576**, A113 (2015). http://doi.org/10.1051/0004-6361/201425088
F. van Leeuwen, A&A **474**, 653 (2007). http://doi.org/10.1051/0004-6361:20078357

Part III
Summary and Future Work

Chapter 7
Summary and Future Work

The light element lithium allows to study several different processes in astrophysics. It is one of the few elements that is produced immediately after the Big Bang, and it is also used extensively as a proxy for mixing inside stars, being very sensitive to environmental conditions. Although it is produced inside stars by nuclear burning, it is rapidly burned by proton capture inside low-mass stars during the main sequence. When the star enters its post-main sequence evolution, the surface composition changes due to the first dredge up. As Li was greatly depleted in the stellar interior during the main sequence, its surface abundance will be diluted during the deepening of the convective envelope, thus predicting giants with very low surface lithium abundances.

However, it is well documented that Li-rich giants do exist. Some explanations for the enhancement have been proposed, but up to this day this anomaly is not well understood, and the lack of a comprehensive explanation for observations of enriched giants presents an opportunity to study noncanonical physics involved in the evolution of stars. Also, different mechanisms of Li enrichment (Li production inside the star plus an extra-mixing mechanism, accretion from an AGB star, and substellar companion engulfment) produce different observable signatures and may even produce different Li signals in the star. They could also be generating Li-rich giants in different evolutionary stages, or different mass and metallicity domains, and as such, it is extremely important to properly characterize the population of Li-enriched objects.

During my Ph.D. thesis, I have studied theoretically the different possible mechanisms to produce a Li enhancement in giants, which leads to interesting predictions that can be tested observationally. We have analyzed the possible internal production of Li that needs to be coupled with an extra-mixing mechanism that transports the produced light element to the convective envelope where it can be observed. After a detailed literature analysis, we see that proposed mixing mechanisms do not seem to be efficient enough to increase the surface Li abundance (Chap. 2). Thermohaline mixing appears to be too slow, meridional circulation

C. Aguilera-Gómez, *Explaining Lithium Enriched Red Giant Branch Stars*, Springer Theses, https://doi.org/10.1007/978-3-030-02583-0_7

and shear instabilities produced by rotation are inefficient, and magnetic buoyancy cannot explain the most enriched objects. Moreover, all of these extra-mixing mechanisms are expected to develop (or be more efficient, in the case of rotationally induced mixing) after the luminosity function bump. Recent observations indicate that Li-rich giants are not restricted to a specific place in the HR-diagram, and even if some authors have found a concentration of them in evolutionary stages like the clump, or the luminosity function bump, those giants located before the luminosity function bump certainly require a different explanation not provided by extra-mixing.

As internal mechanisms fail to explain the entire population of enriched giants, we explore also the external mechanism of Li enhancement, where the material is accreted from an external source (Chap. 3). A quick search for radial velocity variability in known Li-rich giants indicates that most of these stars do not have binary companions, and have not experienced mass transfer from such a companion in their lifetimes. However, it is still possible that they have engulfed their substellar mass companions. The bulk of this thesis is the development of a model of substellar companion engulfment. We have explored a wide parameter space of stellar and companion properties to understand which subset of observations can be explained by this mechanism and see if there is an ideal population where we can test this scenario. Several interesting conclusions are obtained from our models. We have found that, without the complications of added extra-mixing, only certain substellar mass companions can produce a signature in the star, as more massive objects would not dissolve in the convective envelope of the giant. In particular, this limit in the mass of the companion depends on metallicity, but up to [Fe/H]=0.0, we find that only substellar companions with masses lower than $15\,M_J$ can produce an enhancement in Li. Thus, higher mass brown dwarfs that could have higher Li contents to share with the star, would not enrich the envelope, depositing their material in the radiative interior. By changing a series of parameters, we find that the maximum Li abundance that can be reached by the engulfment of a substellar mass companion is $A(Li) = 2.2$, thus not being able to explain the most enriched object. Nevertheless, this mechanism is still very promising because it can explain enriched giants in regions of the HR diagram where other mechanisms fail, as it is the case for giants after the first dredge-up, but before the luminosity function bump, or other regions where extra-mixing does not operate.

A relevant conclusion to this work, that can help us to design good observational experiments to understand the enrichment mechanism in giants, is that the mass of the giant is very relevant, not only because it changes the maximum abundance after the engulfment, but also because it affects the detectability of the signature, and reveals if there really is noncanonical physics operating inside the star. We have seen that higher mass objects can have very high abundances post first dredge up even when evolving in isolation and that the signature of substellar mass companions would be much harder to detect in these stars. On the contrary, it is ideal to study the process of substellar companion engulfment in stars that have evolved from the Li dip and that leave the main sequence with very low Li abundances, as all detections of Li in this type of stars would indicate the presence of an additional non-standard process.

That being said, the typical criterion used to describe Li-rich giants can be misleading. Lower mass objects and very depleted giants that have gone through engulfment can have abundances lower than A(Li) $= 1.5$ but they are enriched. On the contrary, higher mass objects that have no additional physics would be considered Li-rich by this criterion, even when they are evolving in isolation. Our models can also be used to test the underlying planetary population around stars in mass ranges that are harder to study with planet searching techniques such as radial velocities.

On the other hand, other observational signatures, such as ^{6}Li, have been extensively used to distinguish between mechanisms. We show that this isotope is rapidly burned in stars of higher metallicities. Thus, not finding it does not imply that the star has not engulfed a substellar mass companion. Also, this isotope has been searched usually in stars with very high ^{7}Li abundance because it is difficult to detect, as it manifests itself as an asymmetry in the Li doublet. However, as we have seen, there is a limit to the ^{7}Li abundance that can be produced by engulfment, thus these very enriched objects are not produced by this method, regardless of the presence or absence of ^{6}Li. We stress that detecting ^{6}Li in enriched giants that can be explained by planet accretion can be challenging, but that, depending on the metallicity of the star, it can still be an important indicator when A(^{7}Li) $\lesssim 2.2$.

All of these important details emphasize the study of giants with well-known masses and evolutionary phases, such as those in stellar clusters. Trumpler 20 is an example of such a population, that had 2 outliers with high Li abundances, compared to the rest of the giants in the cluster (Chap. 4). We study these giants finding that, contrary to what was suggested regarding the Li pattern of the cluster, there is no need to invoke extra-mixing in the red giant phase to explain the low abundances in the majority of the stars. Instead, we propose that all giants with lower abundances have a canonical evolution and that the two giants with high abundances can be explained by the accretion of a substellar mass companion.

We also present the Li abundances of a sample of giants that are part of a planet search program, and as such can be interesting to study the population of planets around evolved companions, and their relation to the Li abundance (Chap. 5). We think we have identified at least an enriched object in this sample, but some of the other giants with measured Li could also be enriched. A more in-depth study of the entire sample is needed before we can conclude.

At last, stars that are depleted in Li during the main sequence can be especially interesting to test different enrichment mechanisms without contamination from canonically-evolving giants. By defining the properties, such as mass and metallicity, of these main sequence stars we can follow their evolution to the red giant branch phase.

We study the Li desert by using a homogenized catalog of main sequence and subgiants stars, a region in the plane of Li abundance versus effective temperature that shows a discontinuous separation between high and low Li abundances (Chap. 6). This study does not only present a large homogeneous catalog that can be used to study several astrophysical problems related to Li (e.g., relation between planet-hosting status of the star and Li abundance, chemical enrichment of the galaxy, among others), but also we can characterize the stars in the lower side of the Li desert, that seem to be depleted by some mechanism. We have found that these stars tend to be more evolved and are already subgiants. By trying to identify the origin of these subgiants, we can conclude that they are consistent with having evolved from the Li dip, which would explain their depleted Li abundances. Although there is no consensus on the extra-mixing mechanism acting on the main sequence that depletes the Li abundance in the Li-dip, it is interesting to know where we can find stars of this region once they evolved, and hopefully we can continue to trace their evolution to the red giant branch to study other processes, as Li enrichment.

We are one step closer to understanding the Li enrichment mechanisms in stars, but there is still some work to be done:

1. Our models of planet engulfment can be complemented with additional physics so that they can be applied to other regimes. By adding non-convective extra-mixing, we can understand how Li increases or decreases after the engulfment, and to see if somehow the engulfment of the planet can increase the efficiency of mixing.
2. Modeling the increase in Li and the enhanced rotation of stars after engulfment may also provide interesting insight into the problem, and the correlation between two unusual populations, Li-enriched and fastly rotating giants.
3. For giants in Trumpler 20 and other clusters, it would be ideal to obtain abundances of other elements that can help to distinguish between scenarios, like the carbon isotopic ratio. To do this, new observations are required. Also, measurements of Li abundance in turnoff stars would set the expected Li abundance for giants and could confirm if the 2 objects with high abundance are really enriched.
4. Further analysis is needed for our sample of field giants part of the EXPRESS project. We can correct the Li abundances by NLTE, study the trends with atmospheric parameters, and study the outliers we find in the sample. Moreover, we can measure with the same spectra the carbon isotopic ratio.
5. As we have seen in this work, to find the truly uncommon objects and uncover the underlying evolutionary patterns, knowledge of stellar masses is required. Masses of field giants can be hard to measure, but asteroseismology and the recent *Gaia* DR2 allow to measure this parameter and select stars in interesting mass ranges. By selecting a high mass range we can also study the planetary population around A-dwarfs, while selecting stars of mass ranges corresponding to the Li-dip would be ideal to test mechanisms like planet engulfment. New, exciting possibilities are now available with a large amount of data obtained by different surveys. This study changes what is considered interesting as a Li-enriched giant, so some observational programs can be proposed, such as high-resolution follow-up of APOGEE stars in the region of the Li line, design projects specifically to observe stars below the luminosity function bump or

search for other chemical signatures besides lithium. Also, the relation between rotation and Li-enrichment can be key to better understand the processes inside stars. Combining Kepler and K2 data, providing rotation rates, and spectroscopy may be a different direction for the project that could provide interesting insight.
6. Given that masses are important, studying sample with asteroseismic masses (Kepler, K2, TESS) is a good approach to finding solutions to the Li-enrichment problem.

MIX
Papier aus verantwortungsvollen Quellen
Paper from responsible sources
FSC® C105338

If you have any concerns about our products,
you can contact us on
ProductSafety@springernature.com

In case Publisher is established outside the EU,
the EU authorized representative is:
Springer Nature Customer Service Center GmbH
Europaplatz 3, 69115 Heidelberg, Germany

Printed by Libri Plureos GmbH
in Hamburg, Germany